-791

Energy conservation and public policy

	DATE DUE		

ENERGY CONSERVATION AND PUBLIC POLICY

The American Assembly, *Columbia University*

ENERGY CONSERVATION AND PUBLIC POLICY

Prentice-Hall, Inc., *Englewood Cliffs, New Jersey*

Library of Congress Cataloging in Publication Data
MAIN ENTRY UNDER TITLE:

Energy conservation and public policy.

(A Spectrum book)
At head of title: The American Assembly, Columbia University.
Edited by John C. Sawhill.
Background papers for the 55th American Assembly which met in Nov. 1978, Arden House, Harriman, N.Y.
Includes index.
1. Energy policy—United States—Congresses.
2. Energy conservation—United States—Congresses.
I. Sawhill, John C. (date) II. American Assembly.
HD9502.U52E477 333.7 79-1275
ISBN 0-13-277566-2
ISBN 0-13-277558-1 pbk.

Editorial/production supervision by Betty Neville

Manufacturing buyer: Cathie Lenard

The figures on pp. 13, 15, 16, and 18 are from the following source: Energy Modeling Forum, "Energy and the Economy," EMF Report 1, Stanford University, Stanford, California. They are reprinted by permission of the Energy Modeling Forum.

Figure 5 on p. 20 is reprinted by permission of Pergamon Press Ltd., Oxford, England.

Table 3 on p. 161 is reprinted by permission of the Electric Power Research Institute.

Figure 5 on p. 151 is reprinted by permission of *Power Engineering*.

Figure 2 on p. 146 and Figure 3 on p. 148 are used by permission of the National Electric Reliability Council.

Table 1 on p. 50 and Table 3 on p. 55—which are from pp. 26 and 205 of *How Industrial Societies Use Energy*, a Resources for the Future Book—are used by permission of The Johns Hopkins University Press.

10 9 8 7 6 5 4 3 2 1

PRENTICE-HALL INTERNATIONAL, INC. (*London*)
PRENTICE-HALL OF AUSTRALIA PTY. LIMITED (*Sydney*)
PRENTICE-HALL OF CANADA, LTD. (*Toronto*)
PRENTICE-HALL OF INDIA PRIVATE LIMITED (*New Delhi*)
PRENTICE-HALL OF JAPAN, INC. (*Tokyo*)
PRENTICE-HALL OF SOUTHEAST ASIA PTE. LTD. (*Singapore*)
WHITEHALL BOOKS LIMITED (*Wellington, New Zealand*)

Table of Contents

Preface

The energy problem was dramatized for America in late 1973 when the Arabs declared an oil embargo. The immediate result was anger in some quarters, despair elsewhere—and bewilderment everywhere.

Even now, however, "there is still considerable confusion about the real nature of the energy problem and it is still very difficult to get agreement on solutions." So stated the participants in the Fifty-fifth American Assembly, at Arden House, November, 1978. (Their report in pamphlet may be had from The American Assembly.) They agreed that the United States has sufficient energy supplies to last into the twenty-first century but noted no national consensus on the proper combination of these supplies to provide adequate energy. Nevertheless, they noted, a consensus is emerging that "to use our energy resources more efficiently a broad range of activities is needed."

This book was designed by its editor, President John C. Sawhill of New York University, former Federal Energy Administrator, to lend the insights of professional authority to the discussions at the Arden House Assembly. But it is intended to go beyond that circle to the nation at large to add a sense of clarity and reliability to the national debate on energy. We of The American Assembly are confident that its analyses on energy efficiency and productivity will be of help not only to those who make public policy in its final stages but to the general public who will form the climate of opinion without which policy cannot be well made.

The opinions herein are of course the authors' own, for The American Assembly, a national public affairs forum, takes no position on matters it presents for public discussion. Nor did the Ford Foundation or the Atlantic Richfield Foundation, who generously supported a portion of the energy program, have any hand in the making of this book. We are pleased to offer it as a major contribution to advancing national thinking on the subject of energy conservation, a subject *sine qua non* to our national well-being.

Clifford C. Nelson
President
The American Assembly

John C. Sawhill

Introduction

Since 1973 Americans have been confronted with the growing realization that energy is becoming an increasingly scarce commodity and that we can probably look forward to an extended period of rising energy prices, prolonged environmental conflicts, heavy dependence on foreign countries for oil supplies, endless nuclear debates centered around questions of reactor safety, waste disposal, and weapons proliferation, and a host of other difficult and complex energy issues. Yet, even though many of these issues have remained on the national agenda for some time and been discussed in a wide variety of forums, there is still considerable confusion about whether we really have an "energy problem." As one government official recently put it, "the energy problem is that no one really agrees on what the energy problem is."

Various interest groups define the energy problem in terms of the interests of their particular constituencies. Thus, an environmentalist might define the problem in terms of the environmental costs of using coal to replace oil and natural gas and focus on concerns such as (a) the difficulty of reclaiming strip mined land in arid areas, (b) the social impacts on small western communities of transporting coal in unit trains, and (c) the health hazards of burning coal under utility and industrial boilers. Energy industry executives often characterize the problem as the difficulties of expanding domestic energy production in the face of a maze of federal, state, and local regulatory requirements and argue that these regulations reduce or eliminate the economic incentives for new investment, create a climate of uncertainty in which it is difficult to plan, and stretch out the

JOHN C. SAWHILL, *former Federal Energy Administrator, is President of New York University.*

already long time periods necessary to build energy projects to the point where the expected return on investment becomes marginal or nonexistent. Some see the energy problem primarily as a national security threat arising from the heavy U.S. dependence on imported oil from the politically unstable Middle East; others tend to view it more as a financial problem arising from the alleged difficulties of recycling the large Organization of Petroleum Exporting Countries (OPEC) surplus or of financing the rising external debts of some of the less developed countries. Consumer advocates place the principal focus on the impact of higher energy prices on low-income groups and see the solution to the energy problem as finding ways to offset any redistribution of income which might occur as energy prices rise. And some political pundits, in the face of government inaction on a number of these issues, have gone so far as to characterize the energy problem as one more example of the "crisis of democracy."

In the face of these different perceptions of the problem, it is not surprising that the government has had difficulty in developing and enacting energy programs. It is true that several significant pieces of energy legislation have been passed, but as the authors in this book point out, there still remains much to be done. Clearly, the time has come to begin reconciling some of the divergent views about energy and to develop a coherent framework within which the various elements of the energy issue can be addressed. Fortunately, there is the basis for such a reconciliation. There are certain characteristics of the U.S. energy situation which most of those who have studied energy would agree on, and these areas of agreement can serve as the basis for creating a coherent energy policy.

The principle areas of agreement include the following:

1. Nonrenewable sources of energy such as coal, uranium, oil, and natural gas are exhaustible, and as they become increasingly more difficult to find and produce, the costs of production and, therefore, the price of energy to consumers will have to rise. There is a wide band of uncertainty, however, associated with the size of the energy resource base and the cost of exploiting it, making forecasts of future energy market conditions difficult to pinpoint with any degree of accuracy and subject to a wide degree of variability.
2. Oil imports will comprise a significant portion of U.S. energy supplies for many years into the future, and certain of the Middle Eastern states—notably Saudi Arabia, Iran, Iraq, United Arab Emirates, and Kuwait—will continue to play a major role in determining the price and availability of oil in world markets. The recent discoveries in other parts of the world—Mexico, North Sea, Arctic Islands, South China, etc.—will not result in significant additions to world oil production for at least a decade. Political developments in the Middle East will, therefore, affect future world oil

supplies and prices, contributing to the uncertainty associated with future energy market conditions and the difficulty of forecasting.

3. Many of the options available for expanding domestic energy supplies are expensive, highly capital intensive, and will require long lead times to develop (e.g., nuclear plants). Since government action (or inaction) can shorten or lengthen these lead times, public policy will be an important determinant of the future price and supply of energy.
4. Environmental issues will not be resolved quickly or easily. For example, there is still considerable disagreement about the impact of expanding coal production on human health, and issues related to the burning of coal—such as the impact of increased concentration of CO_2 in the atmosphere —seem unlikely to be resolved for several years.

One other point of agreement among so-called "energy experts" is the desirability of conserving energy or—as the term is used in this book—of improving the efficiency with which we use energy. The authors here have chosen to define efficiency in the economic rather than the technical sense. By energy efficiency improvement, we mean obtaining the same end result (heating a home to a certain temperature, refining a ton of aluminum, etc.) with lower cost energy (or enhancing the end result with the same dollar value of energy). The distinction between economic and technical efficiency is one of measuring energy as an economic cost rather than in physical units.

The case for seeking improvements in energy efficiency is compelling. Greater efficiency can lead to less dependence on uncertain foreign sources for oil and thereby reduce the "security risk," shrink the balance of payments deficit, provide more time to accommodate environmental concerns, etc. Yet, the U.S. has often been criticized as a laggard in developing energy conservation programs. A review of the U.S. energy performance in 1973–77, while inconclusive, is encouraging in some major respects. In this period the U.S. GNP rose 8 percent and industrial production 6 percent while total energy consumption rose only 3 percent and oil and gas consumption actually declined by 1 percent.

European and Japanese energy consumption during the 1973–77 period also grew more slowly than historical energy-GNP ratios would have predicted. With a Gross National Product increase of 7 percent (just slightly less than the U.S. figure), European energy consumption grew only 0.5 percent (compared to 3 percent in the United States). However, the European GNP increase included a much smaller component of industrial production growth (2 percent compared to 6 percent in the United States). Since the industrial sector is so much more energy intensive than the rest of the economy, it is reasonable to conclude that the overall energy conservation performances in Europe and the United States were roughly comparable. In most sectors of their economies, however, the European countries and Japan continue to use energy more efficiently than the United States, but the gap may be narrowing. For example, during this period, gasoline consumption in the major European industrialized countries (except Italy) grew more rapidly than in the United States.

The comparison of changes in oil imports presents the U.S. performance in a less favorable light than the comparison of changes in energy consumption. In 1977, U.S. oil imports accounted for about 25 percent of total U.S. energy use, up from 15 percent in 1973. About 8.5 million barrels per day (50 percent of the oil consumed) was imported in 1977, up from approximately 6 million barrels per day (35 percent of oil consumed) in 1973. It is important to recognize, however, that the 34 percent increase in U.S. oil and gas imports (2.5 million barrels per day) that occurred between 1973 and 1977 was entirely due to declines in domestic oil and gas production; it does not reflect poor U.S. performance in conservation. The comparison between the behavior of European and U.S. oil imports appears adverse to the U.S. because, in contrast to the declines in domestic oil and gas production the United States was experiencing, European domestic production increased by 1.8 million barrels per day between 1973 and 1977 and allowed, thereby, a 1.7 million barrel per day reduction in imports. Japan, with insignificant domestic production throughout the period, absorbed oil imports in 1977 at about the same level as in 1973.

In attempting to formulate policies to foster energy conservation, it is important to recognize that there are costs associated with improving energy efficiency just as there are costs associated with expanding domestic supply. For example, most efficiency improvements such as reinsulating a home or replacing a wet kiln with a dry kiln in a cement plant require large front end investments—they are not free. Some of these investments may not be economically justifiable unless U.S. energy prices rise to world market levels. Thus, in evaluating the policy options available, it is necessary to grapple with the difficult question of how higher energy prices will affect short-term and long-term economic growth.

Other questions which arise in attempting to formulate public policy for energy conservation include (a) the improvement we can expect at different energy price levels, (b) the speed with which these improvements will occur, (c) the institutional barriers to improving energy efficiency and what can be done to remove them, (d) the contribution which research and development programs can make to energy conservation and how such programs should be financed, (e) the role of government (federal, state, and local) in stimulating efficiency improvement. These are among the issues and questions addressed in the various chapters in this book. The purpose is to provide government officials and others interested in public policy with an analytical framework and some guidelines for selecting from among the various options available to improve energy efficiency. The authors also identify and discuss those areas where more (or better quality) data are needed to make intelligent policy choices and what types of additional analysis would be desirable.

One question which remains, however, is why, in the wake of a prolonged oil embargo in the early 1970s and natural gas and coal shortages in the latter part of the decade, so many energy issues still remain unresolved. In part, the failure of the American public and its political leaders to develop an energy policy can be explained by the complexity of the problem and the fact that very little analytical work had been done on energy prior to 1973. But these reasons do not tell the whole story. There appear to be some more fundamental questions involved as well. These questions have made it difficult to develop a consensus on energy policy and thus for Congress to act as rapidly as might have been anticipated given the magnitude of the problem and the attention it has received.

The difficulty seems to be that there is very little consensus in America about the way in which economic growth should take place and the pattern that this growth should follow in the 1980s and 1990s. At issue are such questions as how to measure growth, what priorities should be established, how should these priorities be selected, etc. Are we becoming a more materialistic society or is some other model of the future more consistent with prevailing attitudes and values? Those who appear to be more concerned with goals such as protection of the environment, diffusion of political and economic power, and conservation of natural resources tend to favor one set of policies; and those who assign top priority to goals such as economic growth, stability of the dollar, and America's position of world leadership tend to support different policies. Energy policy debates have often become, as a result, a forum for debating the course of postindustrial American society. Thus, an argument about the management of nuclear waste or the need for a higher excise tax on gasoline may be more a reflection of the underlying values of the participants about where society should be headed than a particular concern for the specific issues involved.

These arguments over basic values which have surfaced in the postembargo energy debates are not alien to Western thinking. The tradition of stewardship, for example, has had its proponents in the past, both in classical and modern times. The view that man has a responsibility for handing over to his descendants a nature made more fruitful by his efforts is not, that is, entirely a contemporary innovation. The etymology of the word nature—derived from the Latin *nascere* meaning "to be born, to come into being"—suggests that man's responsibility to nature is to actualize its potentialities and by this means to perfect it. Aristotle and Aquinas both urged that man should respect and cooperate with nature in the same way that a good artist respects his material. And, in our own time, Herbert Marcuse has criticized man's treatment of nature because we have used it destructively, as distinct from seeking to humanize or perfect it.

This book, however, is not designed to deal with these broader philosophical issues and the value judgments that underlie them. Rather, it is more narrowly focused on improving energy efficiency, describing the adjustments in efficiency which will occur in response to market prices, and outlining some of the public policy options available to supplement and enhance market forces. The central question to which the authors address themselves is what policy options are available to accelerate the natural adjustment of energy markets to a regime of higher energy prices. It is helpful, however, in analyzing specific energy policies to understand that the nature of the debate frequently reflects underlying political, economic, and social considerations which go far beyond the energy issues in question.

It also helps to clarify the issues if energy efficiency and energy production are viewed as alternative investment opportunities. Unfortunately, in the postembargo period, a dichotomy has developed between those who support policies to promote energy conservation and those who support measures to increase domestic supply. Thus, President Ford's energy program was criticized by "conservationists" as tilting too strongly in the direction of supply expansion; President Carter's National Energy Plan was criticized by energy industry leaders and others for emphasizing conservation and providing inadequate incentives to develop domestic supplies. What the critics (and the policy-makers) often fail to realize, however, is that investments—either in energy efficiency improvement or in expanding domestic supplies—should be designed to achieve the same ultimate objectives such as reducing dependence on energy imports and minimizing the impact of rising energy costs on economic growth. Society should optimize (in an economic sense) its use of scarce resources so that those investments which promise the highest rate-of-return are the ones which get funded—assuming, of course, that all costs are accounted for and subsidies are equalized or eliminated. From the policy-makers' point of view, the return on an investment in retrofitting homes to make them more energy efficient should be compared to such supply enhancing investments as offshore drilling rigs, nuclear plants, and solar furnaces. Viewed in this framework, investments to improve energy efficiency are no different from investments in coal mines or geothermal wells. The three are merely alternative ways of reducing oil imports, and the one promising the highest payoff should have first call on society's resources.

This way of viewing alternative energy investments is the basis for much of the analysis in this book. The various authors lay out the concept in more detail in succeeding chapters and, in doing so, describe how our thinking about energy efficiency has evolved since the oil embargo as well as some of the improvements in energy efficiency which the U.S. has made in this period. The first two chapters set the stage for what follows in a discussion of the relationship between energy and the economy. In them, William Hogan and Robert Pindyck treat the critical question of energy prices and economic growth in some depth. Hogan takes a macro approach and attempts to show how rising prices might impact economic growth, and Pindyck looks at the relationship between energy prices and the amounts and kinds of energy consumers will demand at different price levels. The intent in both cases is to improve our understanding of how much the marketplace can be relied upon to stimulate efficiency improvements.

In Chapter 3 Lee Schipper analyzes some of the conservation initiatives which have taken place outside of the United States in an attempt to determine what is technically feasible for this country. His work can yield some important lessons for those concerned with energy efficiency in the United States.

The next three chapters present a careful review of the opportunities for improving energy efficiency in the transportation, industrial, and residential sectors. In each case, the barriers to improving efficiency are identified and some specific public policy options suggested. The authors are unanimous in their conclusions that higher energy prices are the most important stimulant to energy efficiency improvements and sanguine about the improvements which will occur as a result of the price increases which have taken place since 1973.

Douglas Bauer and Alan Hirshberg present the special case of the electric utility industry and describe some of the experimental work being done with electricity prices to level out peak demand and reduce, thereby, the need for new electricity generation and transmission facilities. They also take a look at some of the opportunities for improving the efficiency of converting fuels into electricity and conclude that major short-term improvements will be difficult but that there are significant long-range opportunities.

Christopher Hill and Charles Overby describe some of the technological options for improving energy efficiency through recycling and reuse. They particularly focus on the opportunities for burning solid wastes and estimate the kinds of savings which might be possible as these systems are developed. They also deal with such issues as returnable bottles and the opportunities for recycling in the industrial sector. Denis Hayes makes an interesting case for increased use of solar energy. He reviews the economics of various solar technologies and discusses their potential. His analysis suggests that there may be a larger role for "soft technologies" than previous forecasts have identified.

John Gibbons—in a comprehensive chapter on research and development—points out some of the long-range opportunities for improving energy efficiency and defines a role for the federal government in sponsoring research and development programs to take advantage of these opportunities. He presents some interesting ideas which would, if implemented, significantly increase federal funding for energy conservation research and put it more on a parity with the research being done to increase energy supply. His chapter should be a useful input to policy-makers in Washington concerned with government energy research programs.

The final chapter in the book presents some interesting findings about energy efficiency which have heretofore been neglected in the literature. Robert Reisner analyzes the role of state and local governments in energy conservation, and his chapter includes some of the federal government initiatives which have been undertaken since the oil embargo to strengthen state and local government programs. His findings suggest that state programs will become more important in the future and therefore should receive more attention from policy-makers than they have in the past. He also identifies some of the ways in which state grants programs can be improved and administered more effectively. The book concludes with a short summary of the conclusions and recommendations of the authors and some suggestions for further research.

William W. Hogan

1

Energy and Economic Growth

Energy–Economic Determinism?

United States economic output and energy consumption have experienced similar growth in recent decades. Between 1950 and 1973, the economy grew at 3.6 percent per year while energy consumption grew at 3.4 percent per year. It is natural to see a causal relationship in this pattern: expansion of the economy increases energy consumption, and a plentiful energy supply is a spur to economic growth. However, this pattern is associated with a period of relative energy abundance and low energy prices. In contrast, the common expectation is that future energy supplies will be limited and expensive. This new perception of the energy situation and economic growth has created a call for national action. If energy availability determines economic well-being, a large effort is required now to guarantee that our energy needs are rational and that appropriate steps are taken to meet those needs.

At the root of this national concern is an assessment of the dependence of the economy on energy. This is a complex problem. Energy availability affects every facet of our economy, and energy is used in many different forms. What may be true for the use of electric power in aluminum production need not be true for the use of oil in home heating. Regional differences, the long lead times for major changes in facilities, and the uncertainties of the security of supply contribute to the difficulty of describing the interface between the energy sector and the remainder of

---

WILLIAM W. HOGAN *is professor of political economy and holder of the IBM Chair in Technology and Society at the John F. Kennedy School of Government, Harvard University. In 1975–76 he was Deputy Assistant Administrator for Data and Analysis of the Federal Energy Administration.*

the economy. It is not surprising, therefore, that there is a diversity of opinion about the nature and importance of energy-economic interactions.

There is some evidence that the relationship between energy and economic growth is malleable, but the degree of potential flexibility is disputed. Basic physical laws indicate that energy is required for every activity; thus, if adequate energy is not available, the activity cannot take place. It is from this perspective that the historical link between energy and the economy is cited as evidence that their future growth cannot be separated. In the short run, most would agree, for we must use the equipment and processes now in place, and their range of energy utilization is narrowly restricted. In the longer run, however, new equipment can be purchased, alternate transportation systems designed, the mix of desirable products changed, and new technologies introduced. The same level of economic output might be obtained with a lower level of energy utilization and the quality of life maintained, or even improved, some would say. This perspective is supported by the evidence of different energy utilization patterns, with higher energy prices, in other industrialized nations. Hence, the history of low energy prices may explain the rapid growth of energy demand in the United States.

The potential available to us for decoupling energy and economic growth, i.e., improving energy productivity, is the theme of this book. Detailed examples of the flexibility of energy use, ranging from the adjustment of thermostats to the design of integrated community energy systems, are presented in other chapters. But the abundance of opportunities for changing energy utilization patterns produces a complexity that is a barrier to understanding the net effects of the adjustments to energy scarcity; e.g., Hill and Overby in Chapter 8, to illustrate the unexpected interactions of energy and the economy, describe the potential secondary impacts on employment and industrial output of a simple increase in tire recapping. If we approach the problem gradually, however, the many examples can be put in perspective by developing, first, an aggregate representation of the role of energy in the economy. This aggregate view concentrates on the economic costs of energy scarcity, costs incurred while accommodating to higher energy prices: no credit is given here for reducing energy consumption per se, nor is energy seen as the lever for achieving other social objectives. Other things being equal, a reduction in energy availability reduces our economic choices and, consequently, our economic well-being. But, in the long run, the potential flexibility inherent in our economy may be large and the choices many; in energy policy, the journey, through the short-run transition, may be more important than the destination, the long-run pattern of energy utilization.

Costs of Transition

The Organization of Petroleum Exporting Countries (OPEC) embargo and oil price increases of 1973–74, with the concurrent inflation and recession, provided vivid evidence of the importance of energy in our economy. The immediate increase in payments for imported oil alone amounted to more than $20 billion per year and led to higher prices for other sources of energy, increased unemployment, reduced investment, and lower output, all of which added up to a business recession in this country. There were many forces contributing to the recession, of course, but analysts generally agree that the sudden changes in costs and energy availability were the most important of these forces. The indirect effects of the higher import bills more than doubled the direct effects, reducing real Gross National Product (GNP) by 3 to 5 percent.

These effects are substantial, but the severity of the disruption may be an inherently short-lived phenomenon. To be sure, rapid changes in prices are difficult to manage. Most of the indirect costs of the higher prices may be attributed to our difficulty in perceiving what was happening in 1973–74, through the fog of uncertainty, and the resulting application of insufficient or inappropriate corrective measures. The 30 percent reduction in automobile sales in 1974, for example, may have resulted as much from the Federal Energy Administration's gasoline allocation program as from the actions of OPEC.

The analysis of the 1973–74 experience and speculative investigations of hypothetical future price jumps or supply interruptions indicate, with the advantage of hindsight, that we could do better next time. By one account, a rapid doubling of oil prices in the early 1980s would be expensive, but need not by itself cause another recession. Proper management of the transitory indirect effects surely is not easy; it is at the center of our economic dilemma, requiring the accommodation of inflation and the reduction of unemployment. But it should be understood that the problems are not unique to energy, and the solutions are not likely to be found in energy policy. The key to understanding the role of energy in economic growth, therefore, must be in the analysis of the economic forces intrinsic to energy use over the long run.

The indirect response to higher energy prices cuts two ways. Initially, inflation and unemployment present severe problems for the management of the economy, adding to the economic costs of the more expensive energy. Eventually, however, the higher value of energy stimulates everyone to conserve, reducing the economic burden. It is this latter change in energy use patterns that will dominate the long-term impact of higher energy prices. Despite their importance in the short run, therefore, the long-run analysis abstracts from the problems of inflation and unemployment. Without being complacent about the chances of success, therefore, assume that the economy is managed well; then ask how well we can do in the presence of energy scarcity.

The Elephant and the Rabbit?

The long-run link between energy and the economy has been studied intensively in recent years, and many detailed models have been built to provide insight into the problem. The issue is scarcity, with higher energy prices and the impact on the full employment potential of the economy. None of the models implies that energy is not needed for the economy. If energy prices remain stable, then an increase in economic activity should produce an increase in the demand for energy. This would explain the history of energy and economic growth when the prices were low. Of course, the future growth of energy demand may be less than the historical growth because of lower projections for population increase or a trend toward a disproportionately higher growth of the less energy intensive sectors of the economy. The difficult question is, can energy demand growth be dampened further by higher energy prices without proportional reductions in economic activity?

For simplicity in our analysis, we represent the economy in terms of just two inputs—energy and all other items. The energy input is measured as closely as possible to the point of production, i.e., oil at the wellhead. This does not include the capital and labor needed to refine and distribute the energy. Presumably, if energy consumption is reduced, in the long run this capital and labor could be allocated to other uses; valuing energy at the point of production avoids the double counting. Note in Figure 1 that energy is only a small component of the total U.S. economy. As of 1970, the value of primary energy inputs did not exceed 4 percent of the GNP. The analogy of an elephant-rabbit stew illustrates the implications of this low value share. If the recipe for such a stew calls for just one rabbit (the energy sector) and one elephant (the rest of the economy), won't it still taste very much like elephant stew? If energy prices had not risen after 1970, it is likely that energy demands would have grown at about the same rate as the GNP. The 4 percent ratio would then continue into the future. But what would be the effect if energy costs were to double and yet there was sufficient time given to the economy to adapt? One estimate of the impact may be obtained by assuming a constant recipe. Suppose the rabbit is paid for with part of the stew. Then an additional 4 percent of the stew (GNP) must be allocated to cover the doubling in the cost of the rabbit (energy). In fact, other recipes are available that call for less rabbit and, therefore, lead to lower

Fig. 1. GNP and Energy

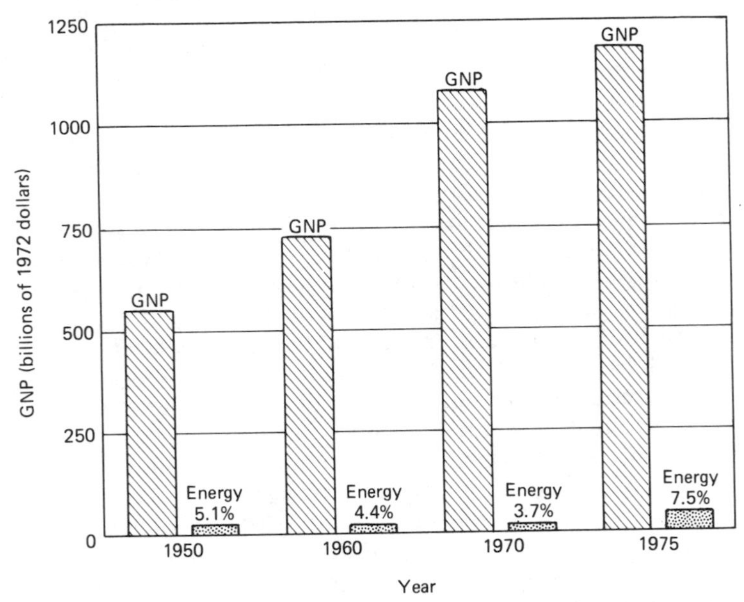

Source: "Energy and the Economy," Energy Modeling Forum, Stanford University.

costs. Under these assumptions, the first doubling of energy costs would produce, at most, a 4 percent loss in GNP.

With a more complicated argument, it can be shown that a small decrease in energy supply leads to a decrease in economic output proportional to the value share of energy in the economy. At a 4 percent value share, a 1 percent reduction in energy input would produce a 0.04 percent drop in total output. By this argument, a small percentage change in energy availability produces a considerably smaller percentage impact upon the economy as a whole.

This simple analysis provides some insight, but it suffers from a major defect in failing to represent accurately the flexibility of energy utilization in the economy. The processes for future production and utilization of energy are not fixed immutably. Within limits, insulation, efficiency improvements, and changes in the mix of input factors can alter the energy requirements for a fixed level of output. Such substitution possibilities can modify the economic impact of changes in the energy system. Flexibility in energy utilization is a central factor in determining energy-economic feedback, and its treatment varies widely among the many different energy models.

Flexibility through Substitution

The specific processes for energy substitution may be varied and intricate. Therefore, even if it is generally agreed that some substitution is feasible, it may not always be possible to identify all the specific technological options available. The many examples of substitution discussed elsewhere in this volume are indicative of the opportunities available, but not exhaustive. The detail may be approached gradually by expanding our first simple analogy based on the value share of energy. We explicitly assume now that substitution is possible between energy and nonenergy inputs to the economy. As energy prices increase, the cost conscious energy user will reduce the use of energy and substitute other resources, such as capital and labor. The end result, after all the many adjustments have been made in the economy, can be summarized by the proportional reduction in energy use, for a given level of output, in response to a change in relative energy prices. To a first approximation, this measure of flexibility can be represented in economists' terms as the elasticity of substitution. Ignoring the feedback to the economy or other inputs, this parameter is the same as the elasticity of energy demand. The elasticity of energy demand measures the percentage response of the demand to a given percentage change in energy prices. Hence, if the elasticity of demand is −0.3, a 10 percent increase in energy prices produces a 3 percent decrease in energy demand, a decrease achieved by substituting capital and labor for energy.

Of course, it is easier to define the aggregate elasticity than it is to identify the many price induced changes in individual energy use patterns or measure the elasticity that these changes imply. The measurement problem is the focus of Chapter 2 by Pindyck, which examines the historical evidence for different economic sectors to provide a range of estimates of the elasticity of demand. Similarly, the detailed conservation examples throughout this volume only indicate the possible future responses to higher energy prices. But, if the measurement problem can be overcome, the aggregate elasticity can be used to examine the impacts of energy scarcity. This concept, the elasticity of substitution, provides a convenient index for summarizing the aggregate behavior of the detailed energy-economic models. If we assume that inputs of other factors such as capital and labor are held constant, then the elasticity of substitution virtually determines the feedback effect of the energy sector on the rest of the economy. The implications of alternate elasticity estimates denoted as σ are shown in Figure 2, which is drawn from a comparative study of energy-economic models. This depicts the GNP in the year 2010 (long enough in the future to ignore transition effects) as a function of energy input, holding other inputs constant. It is assumed that a Btu tax is imposed gradually to reduce energy consumption, with the revenue

Fig. 2. Economic Impacts of Energy Reductions in the Year 2010 for Various Elasticities of Substitution (σ)

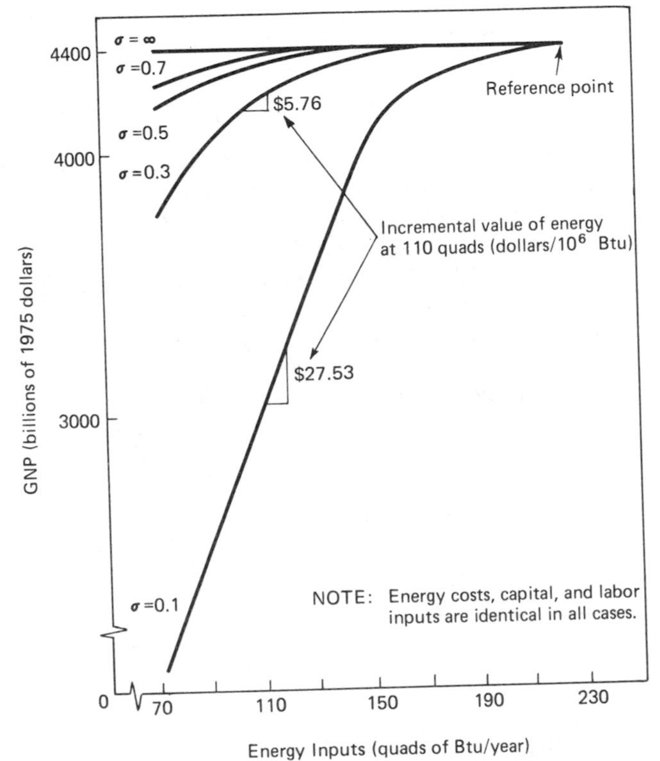

Source: "Energy and the Economy," Energy Modeling Forum, Stanford University.

rebated to consumers. Such a tax might be levied, for example, to mitigate environmental impacts or lessen important vulnerability.

The analysis is based on a hypothetical reference forecast, but the qualitative conclusions are not sensitive to the reference assumptions. A small change in energy availability has almost no effect on GNP. The loss in output is exactly balanced by the savings from the reduced payment for energy. This is what the price represents in a competitive market: the value of the product at the margin. This value will change as the quantity of energy input changes, but the economic output does not decrease in proportion to the decrease in energy input. Substitution of other input factors, such as capital and labor, compensates for the reduction in energy supply.

The importance of the long-run elasticity of substitution is startling in the context of this analysis. A 50 percent reduction in energy availability produces a 28 percent reduction in GNP if the elasticity is as low as 0.1, but only a 1 percent reduction in GNP if the elasticity is as high as 0.[illegible] Seemingly small changes in the substitution potential produce major changes in economic impact. Even the smaller GNP reductions have a large value, however. If the economy is growing at 3 percent in real terms and we discount future consumption at 6 percent, then a 1 percent reduction in annual GNP corresponds to a present value of nearly half a trillion dollars. This is only 1 percent of the present value of future output, but it would justify a substantial research investment aimed at developing low cost technologies which can expand energy supply or improve the efficiency of energy utilization.

An alternative indicator of the economic impact of energy scarcity is found in the implicit tax associated with a given energy reduction. As a measure of the marginal value of energy, this tax may be a more appropriate barometer of the importance of energy scarcity. Although the specific tax is determined by the arbitrary assumptions of the reference forecast, the sensitivity to changes in the elasticity of substitution repeats the results of the analysis of GNP. The implicit tax for the 50 percent energy reduction from the reference forecast is shown in Figure 3. If the elasticity of substitution is as low as 0.1, the necessary tax is $27.53 per

Fig. 3. Implicit Tax, Cutting Energy Use in Half by 2010

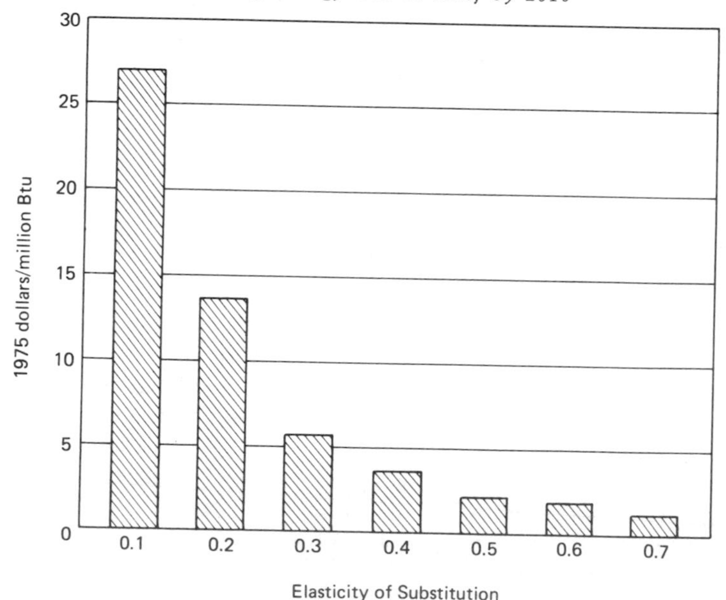

Source: "Energy and the Economy," Energy Modeling Forum, Stanford University.

million Btu's, a tax of over 3400 percent. But if the elasticity of substitution is as high as 0.7, the tax is reduced to $1.26 per million Btu's or 158 percent.

What is the proper elasticity of substitution? As Pindyck explains in detail in Chapter 2, the estimation of this parameter has been the subject of many studies, but there is no definitive resolution of the issue. There are difficulties in comparing definitions, problematical data, and disputes about the relevance of past experience in extrapolations to the future. A consistent interpretation of these studies (adjusted to primary energy prices) indicates the elasticity of substitution is between 0.2 and 0.6, although there is evidence for higher and lower values. The detailed models which have an explicit representation of the full economy yield values between 0.3 and 0.5 for the elasticity as defined here, in terms of primary energy prices. This indicates that there is substantial but not unlimited flexibility in energy utilization in these models. The examples of individual conservation technologies and the empirical evidence cited by Pindyck tend to support the higher estimates of the elasticity.

Capital and Energy

The estimates in Figure 2 of the impact of energy reductions are based on a simplified, partial analysis of the economy. This shows the potential of substitution between economic inputs to absorb energy input reductions with less than proportionate reductions in economic activity. However, changes in energy input have a further dimension of feedback to the economy. This additional dimension centers on the pattern of capital investment over time. Reductions in energy input lead to changes in the rate-of-return on capital as well as reductions in the level of total output. Investment, savings, and capital use are altered as a consequence. Over time, these effects may cumulate into significant changes in the stock of capital and, therefore, in the productive capacity of the economy.

It is difficult to analyze these complex interactions; sophisticated models are required for this task. As a good approximation, however, we can extend the partial analysis to illustrate the magnitude of the economic effects of changes in capital input. For this purpose, expand the beginning framework to include three economic inputs—energy, capital, and labor. Now, instead of holding both capital and labor inputs constant as energy input changes, let capital adjust to maintain a constant rate of return. Since total labor input remains unchanged, the returns to labor are reduced. The impact of this new assumption, for the case of an elasticity of substitution of 0.3, is displayed in Figure 4. For a 50 percent reduction in energy input, there is a 4 percent reduction in GNP when capital is held constant, but the reduction is 11 percent GNP when energy is reduced and capital changes to maintain a constant rate-of-

Fig. 4. Economic Impact of Energy Scarcity in the Year 2010 for Alternate Capital Assumptions (Elasticity of Substitution $\sigma = 0.3$)

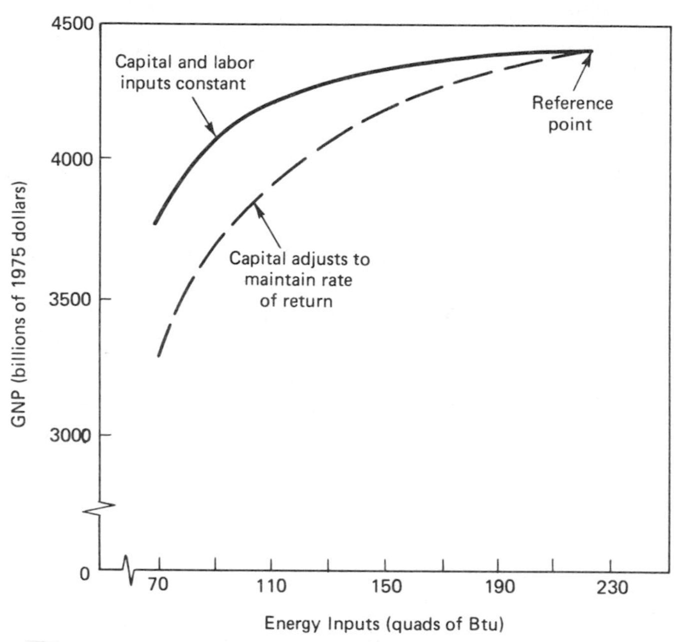

Source: "Energy and the Economy," Energy Modeling Forum, Stanford University.

return. In this case, the capital change effect exceeds the direct effect of the energy reduction. Thus, both substitution and capital adjustment processes are important in considering the feedback effects of energy on the economy. When both these processes are taken into account, however, the conclusion remains: while energy reductions do have a substantial economic impact, the GNP reduction is proportionally smaller than the reduction in energy use and is sensitive to the elasticity of demand.

Distribution of Costs and Benefits

The expectation of increasing energy prices is bad news, but the assessment of the long-run flexibility of our economy is good news indeed. The mitigation of the aggregate economic costs, through substitution and energy conservation, implies that the increasing scarcity of energy need not dominate our future, even though we would like cheaper energy if we could find it. But energy problems are not likely to vanish soon from the center of policy attention. The proportionally small aggregate economic effects disguise potentially large changes in the distribution of energy costs and benefits, and our political system is very concerned with these transfers of wealth. It may be that the attention to energy and aggregate economic growth is misplaced. Perhaps the real issues center on the distribution of energy costs and benefits—across industries, regions, and individuals.

The distributional effects across industries are understood in principle, and several available energy-economy models can represent separately the impacts of energy prices on different industries. Energy intensive industries will be affected most, although less than might be expected. The benefits of substitution and energy conservation are available to dampen the contribution of higher energy prices to the increase of prices for final products and services. Hence, industry can expect to change its mode of operation, but the long-term readjustment of demand should not produce many dramatic changes in industrial composition. The gasoline guzzling automobile industry will be replaced by the gasoline conserving automobile industry and still be healthy. The energy industry itself will see the most pronounced reduction in demand for its products in response to higher energy prices.

The distribution of energy costs and benefits across regions, which may be the most intractable short-term energy policy problem, is not as well understood. We know that the economy, and the use of energy, display a remarkable degree of variation across the continent. There is a range of nine-to-one across regions in the rate of growth of aggregate income and a range of five-to-one across the states in the use of energy per unit of income. The analysis of regional effects is inhibited by the formidable data requirements and by the complex interactions with energy pricing regulations. The energy-economy models do not usually include regional detail, and in the few cases where the attempt is made, there is uncertainty about the supporting data and theory. Yet the conflicts between the energy producing and energy consuming states, the competition for the diminishing supply of price controlled fuels, and the apolitical migration of energy related pollutants cannot be addressed without recognizing the disparities in the regional availability of energy. An improved regional analysis represents a formidable but important analytical challenge. At least one author sees opportunity in this problem: aggressive investment in energy supply may be matched with the regional needs for employment and economic development. At a minimum, we must give increased attention in the future to the important regional differences.

The distribution of energy costs across income classes is a sensitive political problem, often because energy purchases are believed to be a larger fraction of total expenditures for the poor than for the rich. Of course, the rich will always suffer less than the poor from any broadly based income reductions, but the presumption that energy price increases are highly regressive may be inaccurate. It is true that direct energy expenditures, the electricity and gasoline we purchase for our homes and cars, are a larger portion of budgets for the poor than for the rich. But energy use is pervasive in the production of other goods and services; therefore, this energy is used by the consumer, indirectly, through the final purchases of goods and services. When the total purchases of energy, both direct and indirect, are considered, the picture changes substantially. As shown in Figure 5, there is only a slight saturation of total energy purchases for the higher income groups; therefore, an increase in energy prices yields something more like a proportional decrease in income for all groups than a decrease directed primarily at the poor.

The income distribution effects of more selective changes in energy prices are not so clear; e.g., a tax on gasoline might be regressive, but deregulation of natural gas prices might affect homeowners the most. Or the policy concern with distributional effects may stem from a view of energy as a necessity; an increase in energy prices must be accommodated by reducing the consumption of other goods, a reduction which the poor can afford less well than the rich. But this condition is not unique to energy, and there may be better avenues than energy policy for con-

Fig. 5. Energy Purchases versus Household Expenditures

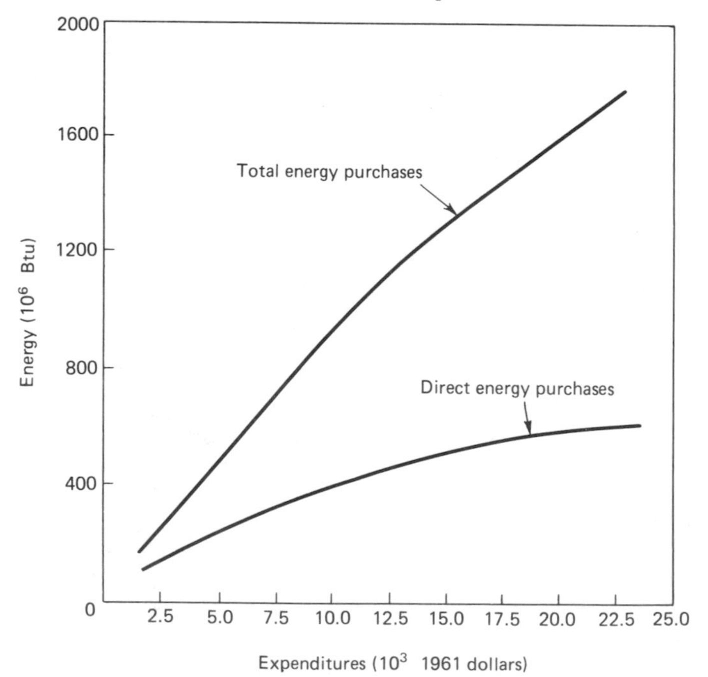

Source: "Energy Cost of Living," *Energy*, June 1976.

trolling the distribution of wealth. There does not appear to be even a simple rule for evaluating the distributional effects of energy policy. The aggregate data indicate that future discussions of the income effects must be related to specific energy policy proposals, and there is a need for better data and models to shed some light on this important dimension of energy policy.

Energy and Economic Growth: Summary

The historical link between energy and economic growth can be explained by the gradual decline in energy prices without adopting the philosophy of energy-economic determinism. In the short run, rapid rises in energy prices are difficult to manage, and there can be severe disruptions because of a sudden change in energy availability. But, in the long run, energy costs are a small component of the economy, and there is a substantial potential for adopting more energy conserving modes of production and consumption; this flexibility can mitigate the aggregate economic impacts of a gradual increase in energy scarcity, whether the scarcity is imposed on us or adopted by us in exchange for other benefits, e.g., environmental improvement. The loss in output is still significant, however, and it may disguise major shifts in the distribution of wealth, shifts that will not be accommodated easily. Energy scarcity is a problem, therefore, but not an insurmountable obstacle. The major difficulties are in moderating the short-run disruptions and compensating for the distributional effects: the journey is more important than the destination.

This diagnosis of the energy-economic growth nexus is optimistic, perhaps excepting comparison with the views of the most enthusiastic conservationists; but it should not be interpreted as a signal to relax. Energy production is one of our most important single industries. Furthermore, the identification of the potential flexibility of energy utilization in the economy is not the same as the realization of that potential. Even the magnitudes are in question. There is a need for a better estimate of the elasticity of energy demand, a topic addressed in greater detail in Chapter 2. And this is only a beginning, for these aggregate estimates of the elasticities summarize but do not identify the essential details of the processes by which the energy substitutions will take place. This is the contribution of the majority of the chapters in this volume: developing process analyses of specific paths for energy conservation. As a first step to achieving this conservation, the energy market must be rationalized, through deregulation and decontrol, so that energy prices reflect the marginal costs of energy production. The higher energy prices will provide the needed incentive; the aggregate analysis indicates that the opportunities are there for both energy conservation and economic growth.

Robert S. Pindyck

2

The Characteristics of the Demand for Energy

From the end of World War II until the early 1970s the price of energy in the United States fell in real (constant dollar) terms, as it did in most other industrialized countries. Between 1950 and 1970 the overall price of energy in the U.S., as measured by a weighted average of the prices of oil, natural gas, coal, and electricity, rose at an average rate of about 0.7 percent per year. The Consumer Price Index, however, rose at an average rate of 2.5 percent per year, so that in real terms the price of energy *fell* by about 1.8 percent per year. Thus, energy in 1970 was about 30 percent cheaper in real terms than in 1950.

Our introductory economics course teaches us that when the price of some good falls relative to other goods, the use of that good should increase. That is indeed what happened with energy consumption. The drop in real prices—combined with an average growth of real Gross National Product (GNP) of about 3.7 percent per year—resulted in energy consumption rising at about 5.5 percent per year, for an increase of about 200 percent over the two decades 1950 to 1970.

Had energy prices continued to fall, we can surmise that energy use would probably have continued to rise at something close to its historic

---

ROBERT S. PINDYCK, *a professor of economics at Sloan School of Management, M.I.T., also holds degrees in electrical engineering. He has written numerous articles in both fields and most recently edited the two volumes* Advances in the Economics of Energy and Resources. *Dr. Pindyck has been consultant to a number of public and private agencies including the Federal Energy Administration.*

rate. The 1970s, however, marked a new era in energy prices. In large part because of the formation of the Organization of Petroleum Exporting Countries (OPEC), but also because of a continually growing scarcity of cheap sources of energy in the United States and elsewhere, the price of energy rose dramatically in the early 1970s. From 1972 through 1977 the price of energy in the U.S. rose about 83 percent in current dollar terms; with a 46 percent increase in the Consumer Price Index over this period, energy prices rose about 37 percent in real terms (just under 7 percent per year, although much of the increase occurred in the years 1973–74). And this increase occurred even though price controls had been in effect for oil and natural gas; prices reached much higher levels in the European countries where, for the most part, controls were less prevalent than they are in the U.S. These price increases have already had an impact on energy demand. The annual rate of growth of energy use in the U.S. fell to about 1 percent for the 1972–77 period, and was actually *negative* in some of the European countries. (This reduction in energy use is only partly due to higher prices; lower rates of economic growth in 1974–75 also served to dampen energy demand.)

How fast energy prices rise in the future will depend in part on the rate at which conventional energy resources become scarcer and more difficult to find, in part on the rate of technological change that lowers the cost of nonconventional energy sources, in part on the behavior of the OPEC cartel, and in part on the domestic energy policy of the United States. Although the dramatic increases of 1973–75 are unlikely, we should probably expect to see energy prices continue to rise in real terms, at least slowly, for the next few decades.

There is little doubt that past and future increases in energy prices will have a dampening effect on future energy demand—as well as at least a temporary dampening effect on employment and economic growth. The questions that are of interest now are first, to *what extent* will higher energy prices reduce energy demand, and second, will higher energy prices (combined with perhaps less energy use) necessarily mean reduced economic growth and a lower standard of living? As we will see, these questions are partly interrelated—both the extent to which prices affect demand *and* the effect of prices on our standard of living depend on the role that energy plays in the production of other goods and as a part of consumers' overall purchases of goods and services.

Answers to these questions are essential to the design of both energy policy and macroeconomic policy. The effects of price changes on energy demand and the extent to which individual fuels can be substituted for each other as their relative prices change will determine the need for and impact of possible energy taxes or other financial incentives as ways of managing the demands for particular fuels. The effects of higher energy prices on the GNP and its growth rate will help determine whether the use of taxes or other energy policy tools might have an undesirable impact on the economy and will also help determine the kinds of macroeconomic policies needed to counteract the possible adverse effects of whatever increases in energy prices do occur.

We have recently seen a considerable amount of public attention focused on energy use in the United States, in part because of the major price increases that have occurred. It is worthwhile to note, however, that oil and natural gas are not the only goods that have experienced sharp price changes. Other commodities have experienced rapid and substantial increases in price—the prices of bauxite and coffee, for example, both tripled in recent years, and the prices of grains and other agricultural products have experienced price fluctuations on the order of 300 or 400 percent over the years. But few people would be as concerned about these events, or expect them to have anywhere near the impact on our standard of living that increases in the prices of energy are likely to have. Is energy somehow a special commodity so that increases in its price must necessarily slow economic growth and reduce our standard of living?

As we will see, energy is indeed a very important commodity, although this does not mean that increases in its price need have a severe impact on the economy. Understanding the impact of higher energy prices requires an understanding of the characteristics of energy demand. It is important to recognize that the impact that higher energy prices will have on our standard of living is a function not just of the *magnitude* of energy expenditures. It is also a function of the particular role that energy plays in our basic consumption patterns and in the production of our goods and services. The impact of higher energy prices will depend on the ability of consumers to use less energy directly and to shift their purchases of goods to those that require less energy, and on the ability of manufacturers to produce their goods using less energy and instead substitute more capital and labor. Let us examine this point in more detail.

Energy Prices and Energy Demand

The impact of a price change on the demand for any particular good can be described by a number called the *price elasticity* of demand. This number is simply the percentage change in the demand for the good resulting from a 1 percent increase in its price. (Similarly, the *income elasticity* of demand is just the percentage change in demand resulting from a 1 percent increase in disposable income or national product, and *cross-price elasticities* give us the effect on demand of a change in the prices of other goods.) If the price elasticity of demand is small, i.e., if demand is price *inelastic*, consumers cannot do without the good and will continue to consume nearly the same quantity even if the price increases significantly. A good whose demand is highly elastic, on the other hand, is one that is less essential and for which other goods can easily be substituted when price rises.

The conventional wisdom, as reflected both in popular opinion and in the working assumptions often used for energy policy analysis, is that the price elasticity of the demand for energy is very small. (Elasticities in the range of 0.2 are often casually suggested as a basis for policy analysis.) The argument behind this conventional wisdom is that increases in energy prices tend to have a much greater impact on consumers and energy using producers than do increases in the prices of other commodities because of the critical role that energy plays, both in the consumption basket and as a factor of production. The argument is made, for example, that consumers have very little flexibility to decrease their use of energy, or even to substitute between alternative fuels, while the consumption of food and other goods can be adjusted much more easily in response to changes in price.

The problem here is that this argument ignores the *time element* inherent in the notion of a demand response to a price change. The demand for a good responds to price and income changes only gradually —and just how gradually depends on the particular good. For this reason we must differentiate between *short-run* and *long-run* elasticities.

It is probably true that in the *short run* energy demand is very price inelastic. If energy prices suddenly increase, consumers cannot, in the space of one or two years, replace their cars with smaller, more fuel efficient ones, replace their energy consuming appliances (refrigerators, air conditioners, etc.) with more energy efficient ones, and insulate their homes and take other measures to significantly reduce their energy consumption. If the price of oil should suddenly rise while the price of natural gas remained fixed (and if supplies of natural gas continued to be available), it is not economical to quickly switch the heating system in one's home from oil to gas, so that the potential for *interfuel substitution* is quite limited in the short run. Similarly, industrial users of energy cannot change their consumption patterns very much in the short run. Most capital equipment was designed to consume a certain amount of energy, so that capital and energy must be used together, i.e., are *complementary* inputs to production. Thus when energy prices increase, producers, at least in the short run, do not have the flexibility to shift to more capital intensive (and less energy intensive) means of production. (Some shift to the use of more labor is possible, but this is costly, since in most industrialized countries labor has become an increasingly expensive factor input.)

But what about the response of energy demand to price changes in the *long run*? The picture might look quite different if, after the price of energy went up, we allowed a considerable amount of time to pass (say seven to fifteen years) before measuring the change in energy consumption. Indeed, while the conventional wisdom about energy demand is probably true for the short run, we will see that it is far from true in the long run, and it would certainly be unfortunate if we based our long-range energy policies on a short-term number for the price elasticity of demand.

An understanding of the true long-run price elasticity is essential to the proper design of energy policy, and to the assessment of the impact of higher energy prices on such macroeconomic variables as inflation, employment, and economic growth. If the household demand for energy is indeed very responsive to price in the long run, eventually the impact of higher energy prices on consumers' budgets will be reduced as the quantities of energy consumed are reduced. In this case deregulation or tax policies designed to reduce or limit household energy consumption through higher prices would be very likely to succeed. Similarly, if energy and capital (and labor) are substitutable in the long run, and if the long-run price elasticity of industrial energy demand is large, then (as William Hogan explained in the preceding chapter) increases in the price of energy will tend to increase the cost of manufactured output by a smaller amount and therefore have a smaller impact on our economy and standard of living.

Even without the use of detailed statistical analyses, there is good reason to believe that energy demand will respond significantly to price changes in the long run. Although significant increases in energy prices occurred only recently in the United States (and not enough time has elapsed to measure the full effects of those price increases), we can look to other countries which, because of their tax policies or for other reasons, have had very different energy prices over the past decade or so. If countries in which energy has always been expensive consume much less energy than countries where energy has been cheap, this might indicate a high degree of price responsiveness. Indeed if we look at gasoline (the demand for which has often been argued to be very inelastic because of its "essential" nature), we find (see Figure 1) that in countries where the price has been high, consumption has been low. We will return to this point later when we discuss the actual measurement of energy demand elasticities.

So far we have talked only about the demand for *aggregate* energy use, but we must also consider the demands for *individual fuels*, i.e., the extent to which individual fuels can be substituted for each other in the long run. Over the next two or three decades, reserves of oil and natural gas may be reduced considerably, so that the availability of moderately priced energy will depend in part on the ability of electric utilities and industrial consumers of energy to switch from these fuels to coal or per

Fig. 1. Cross-Country Comparison of Motor Gasoline Demand in 1975

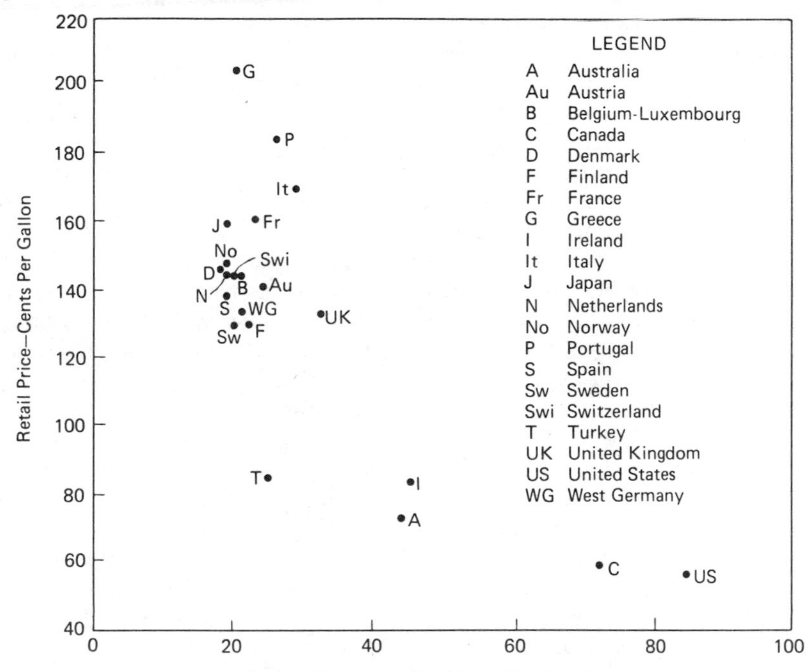

Source: 1978 Economic Report of the President.

haps nuclear power. In the somewhat shorter term, the impact on oil demand of increases in natural gas prices in the U.S. will depend on the extent to which these fuels can be substituted for each other in different sectors. Finally, the extent of interfuel substitutability determines (in part) the impact of an increase in the price of oil or natural gas on the cost of manufactured output, and the impact of a shortage of oil or natural gas on the level of output. Thus a better understanding of the extent of interfuel substitutability and the magnitudes of cross-price elasticities of fuel demands is needed if we are to be able to intelligently design an effective energy policy.

The Structure of Energy Demand

The ratio of energy demand to GNP in the United States rose only slightly during the 1950s and has been fairly constant from 1960 to 1974, leading some people to believe that a more or less fixed proportionality between energy use and total economic output must always hold. Such a belief, however, is completely unfounded. As we explained earlier, up until about 1973 energy prices in the U.S. declined slowly but steadily in real terms, while recently we have experienced large increases in energy prices, increases which may significantly alter the amount of energy used per dollar of output. In addition, in past years energy use per dollar of gross output has varied considerably across countries, and for many countries the ratio has changed significantly over time. This can be seen in Figure 2, where we have plotted energy consumption (in thousands of Btu's) per dollar of Gross Domestic Product (measured in constant 1970 U.S. dollars) for the U.S., Canada, the United Kingdom, the Netherlands, France, and West Germany. Note that the four European countries have energy/GDP ratios well below that of the U.S. and Canada, the ratio for the U.K. has declined somewhat over time, and that for the Netherlands has increased over time. There would certainly seem to be no magic number for the energy/GDP ratio.

Why do we observe these differences in energy/GDP ratios across countries, and why have the ratios increased in some countries, decreased in others, and remained more or less level in still others? An answer often given to this question is that there are differences in life styles across countries which result in different "needs" for energy. Examples often cited for such differences in life styles include different sizes of cars driven because of basic differences in "tastes," or differences in the extent of home heating because of cultural differences or different "habits." While tastes and habits may indeed differ across countries (and across time in any one country), this does not provide a meaningful explanation for the differences that we observe in energy use, and in particular does

Fig. 2. Energy Use per Dollar of Gross Domestic Product

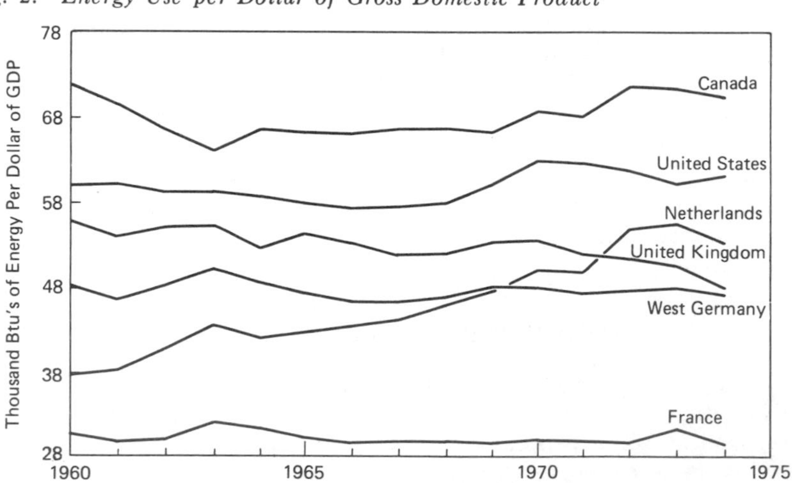

not provide a basis for predicting the kinds of changes in energy use that are likely to come about in the future. We must remember that tastes and habits may *themselves* be functions of price. Taking the price of gasoline and its relationship to average car size as an example, we must ask to what extent car size differences can be attributed to differences in gasoline *prices* across countries and over time. It may well be that people choose to buy smaller and more fuel efficient cars when gasoline prices are higher—in fact, get into the *habit* of driving smaller cars—for *economic* reasons. If this is indeed the case, it has important implications for the impact of higher gasoline prices on gasoline demand in the U.S. and elsewhere.

It is also important to recognize that differences in energy use cannot be explained in the aggregate, but must be explained on a sector-by-sector basis. Clearly the structure of energy demand for home heating will be very different from that for industrial production, so that to look at energy/GDP ratios in the aggregate provides little in the way of useful information. Let us therefore examine briefly the characteristics of energy demand for some individual sectors of use, in particular, the residential, industrial, and transportation sectors.

For the residential (household) sector, the structure of energy demand depends on consumers' relative preferences for energy and other goods, i.e., the willingness of consumers to substitute between energy and other goods in their consumption baskets. For consumers, substituting away from energy as energy prices rise means less direct use of energy (e.g., for home heating and cooling) through better home insulation and through lower thermostat settings in the winter and higher ones in the summer, as well as a reduction in purchases of energy consuming appliances or the replacement of existing appliances with those that are more energy efficient. The conventional wisdom holds that consumers have very strong and inflexible preferences for energy and would be willing to sacrifice large amounts of other goods in order to maintain their use of energy. As we will see, however, this conventional wisdom is not supported by recent statistical evidence.

The characteristics of consumers' preferences also determine whether the consumption of energy (and the consumption of other items) rises proportionately with income growth, or whether income growth, with the prices of all goods held fixed, is by itself likely to produce shifts in the proportions of expenditures allotted to energy and other categories of consumption. Such a shift might occur, for example, if rising incomes encourage a more-than-proportional increase in energy intensive consumption (through, say, increased purchases of labor saving appliances). Finally, we should note that the extent to which fuels will be substitutable with each other as their relative prices change (given some overall level of energy use) is also likely to be different in the residential sector than in other sectors and will depend on the extent to which consumers prefer certain fuels for intrinsic qualities such as cleanliness, security of supply, etc., as well as the capital cost (in the long run) of substituting alternative fuel-burning appliances.

In the transportation sector, the demand for energy will depend on the demand for the specific form of transportation itself and the share in the cost of the transportation service represented by the cost of energy. (If energy costs are only a small share of the cost of the transportation service, then increases in the price of energy will only make a small change in the price of the service and hence only have a small impact on the demand for the service even if that demand is highly elastic with respect to changes in the price of the service itself.) In addition, the demand for energy will depend on the ability to adapt the particular form of transportation to make it more fuel efficient (for example, by building smaller cars or by driving existing cars at slower speeds). The demand for energy in the transportation sector will, of course, vary considerably across particular forms of transportation, because of differences in the demands for the alternative transportation services themselves, differences in the energy cost shares for the services, and differences in the cost of improving fuel efficiency.

For the industrial sector, the structure of energy demand depends on the characteristics of our production processes, and in particular the extent to which capital, labor, and energy can be used in different proportions in response to changes in the prices of these factors. The substitutability of capital, labor, and energy is a critical determinant of the industrial demand for energy; if capital and labor can easily be substituted for energy, industrial producers can reduce their use of energy if its price rises. The characteristics of production also determine whether the industrial demand for energy and the demands for other factors rise proportionally with the growth of industrial output, or whether output growth, with prices of factors held fixed, will by itself produce shifts in the proportions of expenditures allotted to each factor. Finally, the characteristics of production determine the extent to which individual fuels will be substituted for each other as their relative prices change.

Energy demand in the industrial sector, and in particular the substitutability of capital, labor, and energy, is especially important because it determines the *macroeconomic* impact of changes in energy prices. It is important to recognize that the causal relationship between energy and the macroeconomy runs in both directions. Most people are aware of how increases in energy demand are brought about by growth in GNP, but only recently have people become aware of the importance of energy to GNP growth itself. A physical *shortage* of energy (or for that matter any other input used for production) can obviously depress GNP and increase unemployment by creating bottlenecks in the production of both intermediate and final goods. An increase in the *price* of energy, however, can also reduce the productive capacity of the economy. If energy or any other factor of production becomes more scarce (i.e., more costly), this necessarily reduces the production possibilities of the economy, so that GNP will be lower than it would if energy prices had not increased. The question, of course, is *how much* lower GNP will be as a result of an increase in energy prices. As discussed in some detail by William Hogan in the preceding chapter, this depends on the elasticities of substitution between energy and other factors. If the possibilities for substitution are great, then less expensive factors can be used in greater quantity in place of energy.

Because of the important interrelationships between the energy sector and the macroeconomy, energy use—and energy policy—increasingly impinge on the design and impact of economic policy. We are beginning to realize, for example, that the rate of unemployment may depend not only on the particular monetary and fiscal policy in effect, but also on the changes in energy prices and energy use that took place over the past two or three decades.

To see this, consider the fact that between the end of World War II and 1972 a slow but steady shift occurred in the structure of industrial production in the United States and in most of the other advanced economies. During this period two factor inputs of production—energy and capital—became significantly cheaper in real terms relative to a third important input, labor. This shift in relative prices occurred for a number of reasons. Reserves of energy resources, and energy production, were increasing worldwide, which drove down the real cost of energy. Tax policies in many countries (e.g., the investment tax credit in the U.S.), designed to encourage new capital investment as a spur to economic expansion, helped to reduce the growth in the price of capital services. Finally, tax and social welfare policies, combined with greater wage demand on the part of workers, tended to greatly increase the effective cost of labor services for production. This is illustrated for the United States in Figure 3, which shows real price indices for capital, labor, and energy over a decade and a half. We might point out that the picture would look very much the same for Canada and one of the industrialized European countries.

The result of these changing prices was a shift in the factor mix used in production. Gradually, producers replaced labor with less expensive capital and energy. In addition, there is evidence that capital and energy themselves came to be used in a complementary fashion. The particular kinds of machines that made up our capital stock required large amounts of energy to be utilizable, so that there was little or no room for substitution between energy and capital, at least in the short run. (As we will see, however, the evidence indicates that there may be some room

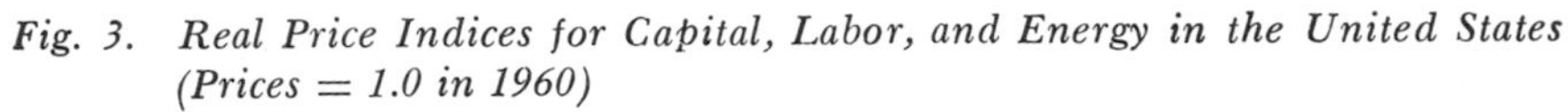

Fig. 3. Real Price Indices for Capital, Labor, and Energy in the United States (Prices = 1.0 in 1960)

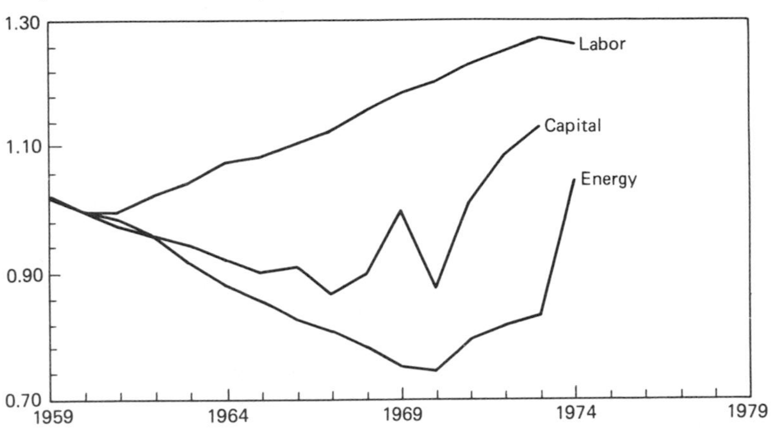

for substitution in the long run, when existing machines can be replaced by new, more energy efficient ones.) This shift in the relative quantities of factor inputs is illustrated in Figure 4, which shows quantity indices for capital, labor, and energy. Observe that energy, which became a relatively cheap factor input, was soon used in larger quantities relative to labor, the more expensive input.

This shift away from labor and toward energy and capital helped to exacerbate the impact of the increases in energy prices that were brought

Fig. 4. Quantity Indices for Capital, Labor, and Energy in the United States (quantities = 1.0 in 1960)

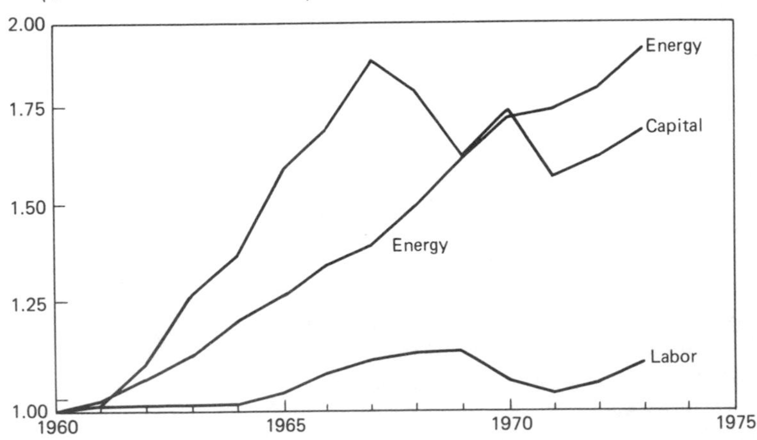

about by the OPEC cartel. When energy prices rose, industries in many countries were unable to achieve a significant shift away from energy intensive production. For at least the short term, energy and capital were complementary inputs, and the only substitutable alternative—labor—was already very expensive. Thus, increases in energy prices were translated into an increase in the cost of industrial output—an increase in cost nearly as large as the percentage increase in the price of energy times energy's share in the total cost of output. But with labor and capital fixed, this had to mean a drop in the level of real output. The result was a recession and the likelihood of lower economic growth during the next decade. In the short run at least, the shift toward more energy intensive and capital intensive production meant a greater reduction in the productive capacity of the economy as a result of higher energy prices. The shift itself, on the other hand, came about because of gradual changes in energy prices (as well as changes in the prices of other factors).

Even if energy prices do not rise very rapidly in the future, the large and dramatic increases in energy prices that have already occurred will have the effect of reducing economic growth in most of the industrialized countries, at least through 1985. The question now is the *extent* to which growth will be diminished over this intermediate range. As we explained earlier, the answer depends in part on the degree of substitutability of capital, labor, and energy in the long run. If capital and energy are substitutable in the long run, then the impact of higher energy prices on the cost of output (and thus on GNP) will be ameliorated somewhat. As we will see, the statistical evidence so far on capital-energy substitutability indicates that there may indeed be a potential for substitution, but only after a considerable amount of time is allowed to pass.

Measuring Energy Demand Elasticities

Economists have for a long time been concerned with the measurement of the characteristics of demand for various goods. Econometric technique (i.e., the use of statistics for the measurement and testing of economic relationships) has been applied to the estimation of own-price and cross-price elasticities and income elasticities of demand for almost every category of consumption expenditure for which reasonable data exist. Over the past several years particular attention has been focused on the econometric estimation of demand elasticities for energy.

Typically, the econometric estimation of the demand for some good begins with the specification of a theoretically sound relationship between the quantity consumed of the good, its price, the prices of other related goods, and some variable such as income or GNP that measures total purchasing power. Although this relationship specifies which variables affect the demand for the good (e.g., the prices of *which* related goods), it does not specify the *magnitudes* of the effects of these variables (i.e., *how much* a 10 percent change in a price will affect demand). To determine these magnitudes (and thus the various elasticities of demand), the relationship must be compared against historical data. This is where the use of statistics comes in. By fitting the relationship to past data on demand, prices, and income, statistical estimates of the demand elasticities can be obtained (estimates that are, of course, only as good as the data used to generate them).

Usually the statistical estimation of a demand relationship is done using *time-series data*, i.e., data over some historical time horizon (e.g., the last twenty years). The idea is that changes in prices and income that occurred in the past can be compared with the changes in demand that followed (and that presumably could be attributed to those changes in prices and income). This approach, which in fact characterized most of the earlier econometric studies of energy demand, may unfortunately have certain shortcomings. First, some of the demand, price, or income variables under study may not have changed very much over the historical time horizon for which data are available. Second, unless the historical time horizon is very long, the approach is likely to capture *short-run* rather than long-run elasticities. The reason is simply that the measurement of successive changes in demand that follow successive changes in prices or income precludes the possibility that enough time had elapsed for a full adjustment to those prices and income changes to have taken place.

These problems—the lack of sufficient variation in the data and the capturing of short-run responses—are likely to result in elasticity estimates that are much lower than the true elasticities (or at least the true long-run elasticities). This may well be the reason why in the past a number of econometric studies of energy demand for a single country (usually the U.S.) came up with small price elasticities.

In order to estimate long-run elasticities of demand, it is necessary to compare the *equilibrium* demands for energy corresponding to prices that are significantly different from each other. By "equilibrium" demand we mean the demand that would prevail after sufficient time had elapsed for consumers to completely adapt to a new price or set of prices. How much time is "sufficient" will depend on the particular sector or subsector of energy use, but it might be anywhere from five to twenty years. Given the limited time horizon for which data are available for any one country, it is unlikely that we can compare equilibrium prices and demands by using data for only a single country. On the other hand, energy prices are and have been quite different *across countries*, so that by using data that span a number of countries, we can indeed compare long-run equilibrium values of energy prices and demands.

The variation of prices and demands across countries and through time is illustrated for the residential sector in Figures 5 and 6. Figure 5 shows a real price index for energy (all prices in constant dollar terms and relative to a price of 1.0 in the United States in 1970) for the U.S., the Netherlands, West Germany, Canada, and France. Note that these prices have declined over time in all countries, but only slowly. The prices, however, vary considerably across countries, with energy prices in West Germany (the highest) about triple those in Canada (the lowest). Figure 6 shows per capita energy consumption in the residential sectors of each of these countries. These levels of energy consumption increase slowly over time, but the greatest variation is across countries so that they

Fig. 5. Energy Price Index for the Residential Sector (Price = 1.0 for U.S. in 1970)

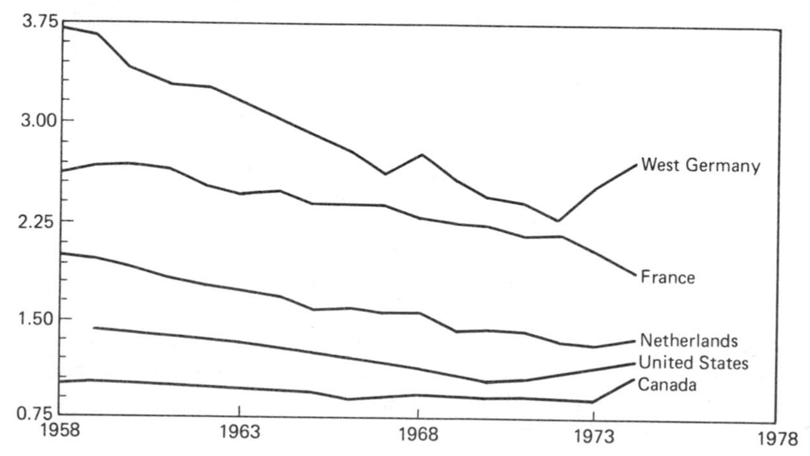

Fig. 6. Per Capita Residential Energy Consumption

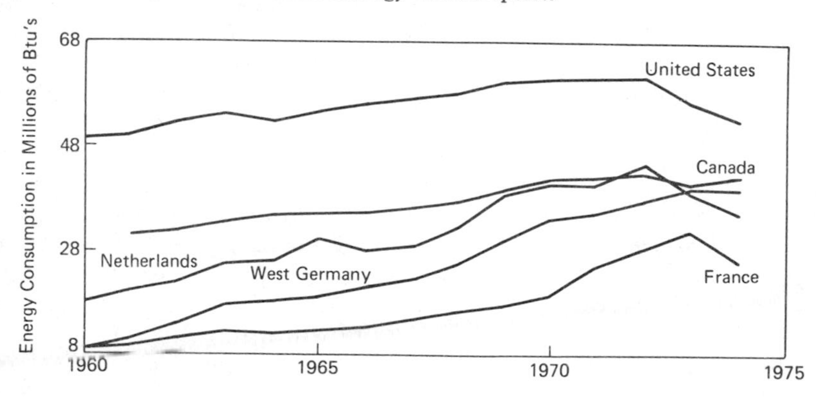

can be viewed as comparative equilibrium levels. (Note that demand is determined by per capita income as well as price. Thus, Canada, which has had the lowest price of energy of the five countries, has a per capita energy consumption below that of the U.S. because its per capita income has been well below that of the U.S. Also, as Schipper points out in Chapter 3, structural differences across countries may partly explain differences in energy use.)

Recently a few studies have used data that span a number of countries in an effort to obtain enough variation in what could be considered equilibrium values of prices and demand levels to elicit long-run elasticities. As might be expected, the results of these studies show much larger values for energy demand elasticities than had been the case with studies using data for only a single country. In particular, a study conducted by this author, and a study conducted by James Griffin of the University of Houston, indicate that long-run price elasticities of energy demand are about twice as high as had been expected previously. We will examine these studies and others next in an effort to evaluate the current evidence on energy demand elasticities.

The Statistical Evidence on Energy Demand Elasticities

In Table 1 we present a survey of recent statistical studies of energy demand elasticities. This survey is by no means exhaustive; numerous other studies have been conducted, but those in the table are first, more recent, and second, give a good overall representation of the existing econometric work on energy demand. The studies and estimates cited in the table are grouped by sector of energy use (remember that the characteristics of energy demand differ considerably across sectors) and by type of elasticity within each sector.

A casual scanning of this table might lead one to conclude that there is wide disagreement about energy demand elasticities, and that it would be impossible to reach a consensus on a set of "working" estimates that could be used for cross analysis. Indeed, looking at the own-price elasticity of aggregate energy use in the residential sector, the long-run estimates range from −0.28 to −1.10. However, let us examine this evidence more closely.

Let us begin with the own-price elasticity of aggregate energy use in the residential sector. The first five studies cited in Table 1 (three for the United States, one for Canada, and one for Norway) actually show rather close agreement on the estimate for this elasticity; all of the estimates are within range of about −0.3 to −0.5. (We will argue that these estimates are too low—although it is worth noting that they are above the kinds of "consensus" estimates of this long-run elasticity that have been applied to policy analysis.) The last two studies cited, however, indicate an own-price elasticity in the range of –0.7 to –1.1. Why have these two studies produced elasticity estimates that are so much larger (about twice as large) than the others?

The first five studies cited in Table 1 are all based on the use of time-series data for a single country. Recall our earlier discussion where we explained that the use of such data is likely to result in *short-run* rather than long-run estimates of elasticities. Because energy prices and consumption levels have changed only slowly over the historical time horizon to which the data applied and because there is insufficient time during this horizon for the full adjustment to price changes to take place, it is unlikely that the use of such data could yield long-run elasticities.

The studies by Nordhaus and this author, however, were both based on cross-sectional data over a number of different countries. Because energy prices and per capita consumption levels differed considerably across countries, it is more likely that these studies would have captured long-run elasticities. Thus, while it is debatable whether the true long-run own-price elasticity of aggregate energy use in the residential sector is closer to –0.7 or –1.1, it is quite likely that it is somewhere within this range and about two or three times as large as the consensus estimates that people have used earlier.

There is little disagreement about the income elasticity of energy demand in the residential sector. With the exception of the Joskow-Baughman study and the Nelson study, all of the other studies show estimates in the range of 1.0. The lower estimates obtained by Joskow-Baughman and Nelson might have again resulted from the use of time-series data.

Estimates for price elasticities of demand for individual fuels used in the residential sector would seem to indicate that those for oil and natural gas are at least –1.0 and possibly larger, so that there should be considerable room for interfuel substitution between these fuels. There seems to be more disagreement about the elasticity for electricity demand, with estimates varying between –0.3 and –1.2. One would expect, however, that electricity demand would be less elastic than fuel oil or natural gas demands since for many uses (e.g., household appliances) there is simply no substitute for electricity.

Let us now turn to the industrial sector and the elasticities of demand for energy as well as other factors of production. Three of the studies cited in Table 1 used aggregate time-series data for a single country: Berndt-Wood (the U.S.), Fuss (Canada), and Magnus (the Netherlands). The first two of these find own-price elasticities for energy demand similar to those obtained from time-series studies of the residential sector, about –0.49. In addition, they both indicate that capital and energy are *complementary*, rather than substitutable. (Note that the cross-price elasticity of capital and energy is negative, so that an increase in the price of energy results not only in the use of less energy, but also in the

TABLE 1. ALTERNATIVE ESTIMATES OF ENERGY DEMAND ELASTICITIES (All Estimates for the United States Unless Otherwise Indicated)

Elasticity	*Study*	*Estimate*
	A. Residential Sector	
Aggregate energy use—own-price elasticity	Joskow and Baughman [1]	S.R.: −0.12, L.R.: −0.50
	Nelson [2]	−0.28
	Jorgenson [3]	−0.40
	Fuss and Waverman [4] (Canada)	−0.33 to −0.56
	Rødseth and Strøm [5] (Norway)	−0.30
	Nordhaus [6] (6 countries pooled)	−0.71
	Pindyck [7] (9 countries pooled)	−1.10
Aggregate energy use—income elasticity	Joskow and Baughman [1]	S.R.: 0.10, L.R.: 0.60
	Nelson [2]	0.27
	Fuss and Waverman [4] (Canada)	0.83 to 1.26
	Rødseth and Strøm [5] (Norway)	1.08
	Nordhaus [6] (6 countries pooled)	1.09
	Pindyck [7] (9 countries pooled)	1.0
Fuel consumption—own-price elasticity	Joskow and Baughman [1]	Gas, oil, coal: all between −1.0 and −1.1 in long run
	Halvorsen [8]	Electricity: −1.0 to −1.2
	Liew [9]	Natural gas: −1.28 to −1.77; Electricity: −0.40
	Hirst, Lin, and Cope [10]	Gas, oil, and electricity: all between −0.84 and −0.91
	Pindyck [7] (9 countries pooled)	Oil: −1.1 to −1.3, gas: −1.3 to −2.1, electricity: −0.3 to −0.7
	B. Industrial Sector	
Price elasticities for factor inputs	Berndt and Wood [11]	η_{KK}: −0.44, η_{LL}: −0.45, η_{EE}: −0.49, η_{KE}: −0.15, η_{LE}: 0.03
	Halvorsen and Ford [12] (2-digit industries)	η_{KK}: −0.67 to −1.16, η_{LL}: −0.28 to −1.55, η_{EE}: −0.66 to −2.56
K = capital	Fuss [13] (Canada)	η_{KK}: −0.76, η_{LL}: −0.49, η_{EE}: −0.49, η_{KE}: −0.05, η_{LE}: 0.55
L = labor E = energy	Magnus [14] (Netherlands)	η_{KK}: 0.05, η_{LL}: −0.26, η_{EE}: −0.90
	Griffin and Gregory [15] (9 countries pooled)	η_{KK}: −0.28, η_{LL}: −0.20, η_{EE}: −0.80, η_{KE}: 0.13 η_{LE}: 0.11
	Pindyck [16,7] (10 countries pooled)	η_{KK}: −0.29 to −0.78, η_{LL}: −0.23 to −0.66 η_{EE}: −0.83 to −0.85, η_{KE}: 0.02 to 0.08 η_{LE}: 0.0 to 0.09

Table footnotes are on page 40.

Elasticity	Study	Estimate
Fuel consumption—own-price elasticity	Halvorsen [17]	Oil: −2.82, gas: −1.47, coal: −1.52, electricity: −0.92
	Fuss [13] (Canada)	Oil: −1.30, gas: −1.30, coal: −0.48, electricity: −0.74
	Pindyck [16,7] (10 countries pooled)	Oil: −0.2 to −1.2, gas: −0.4 to −2.3, coal: −1.3 to −2.2, electricity: −0.5 to −0.6
	C. Transportation Sector	
Motor gasoline, own-price elasticity	Houthakker et al.[18]	S.R.: −0.07, L.R.: −0.24
	Ramsey, Rasche, and Allen [19]	−0.70
	Fuss and Waverman [4] (Canada)	−0.22 to −0.45
	Adams, Graham, and Griffin [20] (20 countries pooled)	−0.92
	Pindyck [7] (11 countries pooled)	−1.3
Motor gasoline, income elasticity	Houthakker et al.[18]	S.R.: 0.30, L.R.: 0.98
	Ramsey, Rasche, and Allen [19]	1.15
	Adams, Graham, and Griffin [20] (20 countries pooled)	0.54
	Pindyck [7] (11 countries pooled)	0.8
	D. Electric Generation Sector	
Electricity generation, fuel price elasticities	Griffin [21]	Oil: −1.0 to −4.0, gas: −0.8 to −1.2, coal: −0.5 to −0.8, cross-price elasticities between 0.2 and 1.2.
	Atkinson and Halvorsen [22]	Oil: −1.5 to −1.6, gas: −1.4, coal: −0.4 to −1.2, cross-price elasticities betwen 0.4 and 1.0.
	Joskow and Mishkin [23]	Elasticities vary greatly across utilities. Fuel price elasticities will be very large when expected fuel prices are roughly the same, but very small when fuel prices differ considerably.

FOOTNOTES FOR TABLE 1

[1] Joskow, P. L., and M. L. Baughman, "The Future of the U.S. Nuclear Energy Industry," *Bell Journal of Economics,* Vol. 7, No. 1, Spring 1976.

[2] Nelson, J. P., "The Demand for Space Heating Energy," *Review of Economics and Statistics,* November 1975.

[3] Jorgenson, D. W., "Consumer Demand for Energy," Harvard Institute of Economic Research, Discussion Paper 386, November 1974.

[4] Fuss, M., and L. Waverman, "The Demand for Energy in Canada," Working Paper, Institute for Policy Analysis, University of Toronto, 1975.

[5] Rødseth, A., and S. Strøm, "The Demand for Energy in Norwegian Households with Special Emphasis on the Demand for Electricity," University of Oslo, Institute of Economics, Research Memorandum, April 1976.

[6] Nordhaus, W. D., "The Demand for Energy: An International Perspective," unpublished, September 1975.

[7] Pindyck, R. S., *The Structure of World Energy Demand,* M.I.T. Press, March 1979.

[8] Halvorsen, R., "Residential Demand for Electric Energy," *Review of Economics and Statistics,* February 1975.

[9] Liew, C. K., "Measuring the Substitutability of Energy Consumption," unpublished, December 1974.

[10] Hirst, E., W. Lin, and J. Cope, "An Engineering-Economic Model of Residential Energy Use," Oak Ridge National Laboratory, Technical Report #TM-5470, July 1976.

[11] Berndt, E. R., and D. Wood, "Technology, Prices, and the Desired Demand for Energy," *Review of Economics and Statistics,* August 1975.

[12] Halvorsen, R., and J. Ford, "Substitution Among Energy, Capital, and Labor Inputs in U.S. Manufacturing," in R. S. Pindyck, editor, *Advances in the Economics of Energy and Resources,* Vol. I, J.A.I. Press, Greenwich, Conn., 1978.

[13] Fuss, M. A., "The Demand for Energy in Canadian Manufacturing," *Journal of Econometrics,* Vol. 5, 1977.

[14] Magnus, J. R., "Substitution Between Energy and Non-Energy Inputs in the Netherlands: 1950–1974," *International Economic Review,* forthcoming.

[15] Griffin, J. M., and P. R. Gregory, "An Intercountry Translog Model of Energy Substitution Responses," *American Economic Review,* December 1976.

[16] Pindyck, R. S., "Interfuel Substitution and the Industrial Demand for Energy: An International Comparison," *Review of Economics and Statistics,* forthcoming.

[17] Halvorsen, R., "Energy Substitution in U.S. Manufacturing," unpublished, 1976.

[18] Houthakker, H. S., P. K. Verleger, and D. P. Sheehan, "Dynamic Demand Analyses for Gasoline and Residential Electricity," *American Journal of Agricultural Economics,* Vol. 56, No. 2, May 1974.

[19] Ramsey, J., R. Rasche, and B. Allen, "An Analysis of the Private and Commercial Demand for Gasoline," *Review of Economics and Statistics,* November 1975.

[20] Adams, F. G., H. Graham, and J. M. Griffin, "Demand Elasticities for Gasoline: Another View," Discussion Paper No. 279, Department of Economics, University of Pennsylvania, June 1974.

[21] Griffin, J. M., "Interfuel Substitution Possibilities: A Translog Application to Pooled Data," *International Economic Review,* October 1977.

[22] Atkinson, S. E., and R. Halvorsen, "Interfuel Substitution in Steam Electric Power Generation," *Journal of Political Economy,* October 1976.

[23] Joskow, P. L., and F. S. Mishkin, "Electric Utility Fuel Choice Behavior in the United States," *International Economic Review,* October 1977.

use of less capital.) These results are somewhat discouraging, but again, they are based on time-series data for a single country and well may represent the short-run, rather than long-run, characteristics of industrial energy demand.

The studies by Griffin-Gregory and Pindyck are both based on pooled international data. Both of these studies indicate a much larger own-price elasticity for energy demand (–0.8 or greater) as well as *substitutability* between capital and energy (note that the cross-price elasticities between these factors are now positive; so an increase in the price of energy results in some substitution toward capital). These results are encouraging, and, as can be seen from some of the other chapters in this book, it makes considerable sense that capital and energy should be substitutable, *given that enough time is allowed to pass.* One would expect that as machines eventually wear out, they might be replaced with more energy efficient ones if the price of energy has indeed increased significantly.

Finally, it is worth noting the interesting results of the Halvorsen-Ford study. This study estimated elasticities for eight individual industries on the grounds that the potential for factor substitution might vary widely across industries. Indeed, they obtained estimates for the own-price elasticity of energy demand that range from –0.66 to –2.56. This reinforces our belief that energy demand is indeed quite price elastic in the industrial sector, but also indicates that more detailed analyses are needed at the level of individual industries. In fact, not only are the *magnitudes* of energy demand elasticities likely to differ considerably across industries, but because of differences in capital depreciation rates across industries, the *speeds of response* to price changes are likely to differ as well. (The nature of these interindustry differences is discussed in more detail in Chapter 6.)

There seems to be little agreement about price elasticities for individual fuels used in the industrial sector, except that again electricity demand seems to be less elastic (as we would expect) than the demands for other fuels. Otherwise, it would seem difficult to pinpoint single numbers or "consensus" estimates of elasticities for individual fuels. The reason for this is probably that the demands for individual fuels vary considerably across industries; an industry that uses a fuel as a chemical feedstock or for some special application would exhibit almost no flexibility in demand, while industries that use fuel simply to generate steam would probably have considerable flexibility in choosing the fuel they use. Again, detailed industry-specific analyses are needed.

We turn next to the transportation sector and, in particular, the demand for motor gasoline. Again, there seems to be a basic difference between the results obtained through the use of data for a single country (the first three studies) and those that use data spanning a number of countries. The first three studies find an own-price elasticity (long-run) in the range of −0.22 to −0.70. It is in fact the lowest end of this range that has been most widely used in policy analyses involving gasoline taxes or other incentives to reduce demand. Even the high end of this range, however, may be too low. As indicated from the last two studies cited, the use of data that span a number of countries, and that therefore are more likely to pick up long-run elasticities, indicates that this elasticity might be much closer to about −1.0. This would indicate that higher gasoline prices could be quite effective as a means of reducing demand, but that one must wait a number of years for the effects of price increases to take place.

Finally, we turn to the choice of fuels used in the electric generation sector. Here little work has been done, and we can cite only three studies. The first two indicate considerable room for interfuel substitution in electricity generation, which would indicate that financial incentives might indeed be effective as a means of shifting from one fuel to another. These results are in fact reinforced by the third study cited, which indicates that the elasticities vary greatly across utilities, but will be large when expected fuel prices are roughly the same in terms of thermal content. In other words, if fuel prices differ considerably (e.g., natural gas is much cheaper on a Btu basis than oil or coal), the cheap fuel will almost always be chosen, even if minor shifts occur in its price. As one would expect, small price changes will affect the choice of fuel only when the relative fuel prices are already quite close to each other.

In summary, we find that the elasticity of energy demand is a dynamic concept, i.e., it will depend on how much time is allowed to elapse after the price of energy has changed. We find, however, that if enough time is allowed to elapse, the demand response can be considerable, i.e., long-run price elasticities are quite large. How much time is "enough" is unfortunately difficult to say and will depend not only on the sector of use but on the particular use within each sector. Unfortunately most uses of energy involve expensive capital equipment (industrial machinery, home heating equipment, major appliances, and automobiles), and in many cases it is not easy to convert this capital equipment quickly. On the other hand, most of the debate over energy policy is concerned with how best to achieve what are essentially long-run targets. We can thus expect price to be an extremely potent policy instrument.

It is interesting to note that our conclusions about the long-run price elasticity of energy demand also have implications for the future price of energy. The world price of oil is largely determined by the OPEC cartel, and changes in this price tend to drive changes in the prices of other fuels. Thus OPEC has, within limits, the ability to manipulate world energy prices. Of course the prices actually faced by consumers will depend also on the taxes and/or price controls in effect in individual countries, but these prices are still very much a function of OPEC's decisions with regard to the world price of oil. OPEC's ability to raise price, on the other hand, is to a considerable extent dependent on the price responsiveness of total energy demand (as well as the price responsiveness of non-OPEC energy supply). If in the long run energy demand is indeed as price responsive as we now believe it to be, then OPEC cannot increase oil prices very much in the future without incurring significant revenue losses. Thus our assessment of the characteristics of energy demand leads us to expect that in the future energy prices are likely to rise only slowly in real terms.

Energy Demand and Energy Policy in the United States

We have seen that higher prices (through deregulation, taxes, or other measures) can indeed reduce energy demand significantly, but it will take some time, perhaps seven to ten years or even longer, to see the full effect of these price increases. While demand elasticities certainly vary across individual industries, in the aggregate this price responsiveness holds across all sectors of use. If we were to pick "working" estimates for price elasticities of energy demand, we might choose a number on the order of −0.8 to −1.0 for the residential sector, a number like −0.8 for the industrial sector, and a number like −0.9 for gasoline demand. It is interesting to note that all of these numbers are at least twice as large, and in some cases three or four times as large, as those that have often been used previously for policy analysis.

We also find that there seems to be considerable room for interfuel substitution in the residential, industrial, and electric generation sectors, although again particular elasticities may vary considerably across industries and across individual electric utilities. It would appear, therefore, that taxes or other financial incentives to encourage "fuel switching" should work, at least if they are applied to the particular industries where there is indeed room for switching. Here we should stress that individual industry studies are needed so that we can better understand where and how to apply fuel switching incentives.

We found that in the long run energy demand in the industrial sector is indeed reasonably elastic. In addition, capital and energy do appear to be substitutable—but only weakly so, and again only in the long run. This would imply that energy price increases could indeed have a significant macroeconomic impact, but an impact that will be ameliorated after a number of years have passed. To the extent that factor substitutability is limited and takes a considerable amount of time to occur, energy price increases, at least in the short term, will be directly translated into an increase in the cost of industrial output, and, with the available quantities of other factors (labor and capital) limited, this means a reduction in real GNP. There is, however, a ray of hope here, namely that it is possible that econometric studies have not captured and perhaps cannot capture the true extent of factor substitutability in the long run. It may well be that detailed engineering studies are needed to tell us exactly how much capital-energy substitution is possible in each particular industry. As the chapters by Reid and Chiogioji and by Savitz and Hirst make clear, engineering analyses of specific industries are being performed and might in fact determine that there is even more room for capital-energy substitution than we would have thought to be the case from the econometric and other statistical evidence.

We should also stress the evidence that exists showing that the price elasticity of gasoline demand is quite large, much larger than the conventional wisdom had held to be the case. Statistical studies refute the conventional wisdom here and show that increases in gasoline prices can indeed significantly reduce gasoline consumption, again allowing sufficient time to pass for a turnover in the stock of cars to take place.

To the extent that energy conservation is a goal of energy policy, it is clear that the most effective policy instrument is the price of energy itself. We now have good reason to expect that our consumption of energy will indeed be reduced if the price of energy is allowed to rise. Other governmental policies (such as building codes, tax credits for home insulation, fuel efficiency requirements for cars, fifty-five miles per hour speed limit) serve to accelerate the response of demand to price changes, but they are unlikely to have a significant impact on demand by themselves. If we are really serious about limiting our consumption of energy, then we must be ready to allow prices to move to their market levels.

If the *supply* of energy is also responsive to price (and there is now a growing body of statistical evidence that it is), the most effective energy policy would be one that deregulated prices. Price regulation has been the direct cause of our growing dependence on imported energy. Allowing the price of energy in the United States to climb to the world level (energy in the U.S. is now about 25 percent cheaper than it is in the rest of the world) will effectively reduce our energy consumption, will stimulate domestic production, and will thereby reduce our growing level of imports.

For a long time the goal of low energy prices has dominated our national energy policy. (This is not surprising, since politicians often lose votes when they attempt to raise the price of anything.) We have instituted a crude oil price controls-entitlements program that basically works by taxing the domestic production of oil and using the proceeds of the tax to subsidize imports. We have regulated the wellhead price of natural gas so that for many years it has been far below the world market level, thereby removing the incentive for producers to explore for and extract gas, while encouraging the wasteful consumption of gas (so that shortages finally developed in the early 1970s). And ironically, while we try to keep its price low, the *true cost* of energy to the American public is rising—as it must. Consumers are now beginning to pay for growing imports of liquified natural gas (LNG) that will cost four to five dollars per thousand cubic feet (equivalent to thirty dollars per barrel oil), taxpayers will soon be asked to subsidize noneconomical sources of energy such as gasified coal and shale oil that would otherwise be unnecessary, and we are all forced to pay the cost of the contribution to the rate of inflation from the decreasing international value of the dollar brought about by rising oil imports.

Whatever the price set by American policy, energy is no longer a cheap commodity. Directly or indirectly, Americans will pay more for it. In the long run the cost will be lower if we pay for our energy directly by letting the price mechanism work. Allowing prices to rise to a free market level will indeed help us to curb our energy consumption and raise our "energy productivity." In fact, it is the only really effective policy tool we have!

Lee Schipper

3

Energy Use and Conservation in Industrialized Countries

Introduction

One of the important aspects of America's painful adjustment to energy realities since 1973 has been an overwhelming effort to look carefully at how we use energy. Much to our surprise there was tremendous slack in energy use at home even before the oil embargo, slack that could have been eliminated profitably. One suggestion that there was waste in our economy came from careful inspection of energy use elsewhere.

But discussion of energy use in other lands has been marred by many distortions and misunderstandings, not only on the part of those who tend to doubt the potential for energy conservation, but even among conservation's strongest supporters. This misunderstanding arises from comparisons of energy use and gross national product, two quantities that have charmed energy statisticians for decades. Though serious work cannot be based only on relationships between two such aggregated quantities, it is useful to review them in the context of some of the popular myths surrounding energy comparisons among countries.

Myths about Energy and the Economy

"Other countries use less energy and enjoy the same standard of living: so can we." This statement has been heard often, most promi-

---

LEE SCHIPPER, *a research professor at the Lawrence Berkeley Laboratory, California, spent the academic year 1977–78 at the Swedish Royal Academy of Sciences and was awarded the King's Medal for his work in Swedish and American energy studies.*

nently in President Carter's energy speeches and almost always refers to the well known fact that the use of energy per unit of national income (E/G) is considerably lower in Europe than in the U.S. (and Canada). While much fruitful work has shown that indeed other countries do use energy more efficiently than we in many important uses, the reverse is sometimes true, too. Most important, though, the ratio of energy to GNP is inappropriate as a measure of energy efficiency because effects of climate, economic structure, life style, and the composition of imports and exports can also affect this ratio. We reject the use of gross comparisons either for purposes of justifying increases in energy use or for urging on conservation.

"There is a strong correlation between energy use and GNP, both in time and among countries." This correlation is often used for purposes of energy forecasting. There is no question that correlation between energy and GNP, either cross-sectional in time or for a single country, can be found with good statistical accuracy. Important for the energy debate, however, is that the price of energy is missing from this simplified measure. While energy prices fell for decades, they are certain to rise now. When even the most simple energy-GNP models are rerun using energy prices as an intermediate variable and a variety of couplings (or "elasticities") relating price to level of use (see Pindyck's or Hogan's discussion in this book), the results are usually much different than in the simplified case—the ratio of energy to GNP, for the case of the U.S., can vary by a factor of more than two for a factor of four variation in energy price. Moreover, among countries at any given level of income there is a fairly wide variation in per capita energy use, again indicating that the correlation, while impressive, leaves considerable scatter. The only firm rule that emerges is that in the long run *THERE ARE NO FIRM RULES RELATING ENERGY USE TO GROSS NATIONAL PRODUCT*. Many factors must be considered, none of which are exposed in the simplistic analyses offered by many critics of conservation. We should avoid use of E/G unless no other measures are available.

"There is so much variation in E/G that intracountry comparisons are not interesting." This critique of recent international comparisons overlooks the possibility of comparisons at a very disaggregated level.

"High productivity in the U.S. is due to high energy use per worker." Though machines (capital) that improve productivity also use energy, the overwhelming share of energy is for space and process heating in any advanced economy, not for operating machines. And most technologies that have advanced productivity have also decreased energy requirements per unit of product. That energy use per worker has increased in most countries is easily explained—machines have increased output/worker faster than output/energy because wages tended to rise while energy costs fell. But energy use per worker, especially among different countries or

different industries, is an especially deceptive measure that should not be used casually in discussions of energy conservation.

"Industries that use less energy per worker (or per unit of product), like leather or textiles, are 'more efficient' than those that use more energy, such as plastics." This popular idea, advanced by some environmentalists, has little validity for energy conservation arguments because different products and processes are being compared whose output is used in different ways. It is not clear to what degree (and at what cost) leather could resubstitute for plastic in, say, seat covers. Nor is the allegation that the energy cost of a shirt made of synthetic fibers is greater than that of a "natural" cotton shirt necessarily any grounds for conclusions about energy use—energy, as one of many production factors or resources, is not the only factor that deserves attention in the energy debate and certainly not the only factor in productivity.

"The U.S. uses more energy per capita (or per dollar of GNP) than most countries because":

1. "Our exports are large." *Wrong*—our exports are relatively much smaller than those of most other industrialized countries, and we tend to import a small amount of energy embodied in foreign trade, while most western European countries export embodied energy.
2. "We feed the world." *Wrong*—agriculture, especially grains and other export staples, is not energy intensive on an energy per ton or energy per dollar basis compared with paper, steel, etc., as Reid and Chiogioji point out in Chapter 6.
3. "Our defense is large, especially in other countries." *Wrong*—because defense includes services as well as airplanes, and our overseas forces draw on energy supplies from other countries. Furthermore, our export of high technology and weapons is not energy intensive (on a per dollar basis) compared to our import of raw materials.
4. "Our industry produces more energy intensive products than 'theirs.'" *Wrong*—while Switzerland, as a single example, produces far less heavy industrial output (paper, steel, chemicals) than we, most other European countries and Japan actually produce more per capita than we. Indeed the structure of U.S. or Swiss industry appears to be more evolved beyond energy intensive raw materials toward higher value-added products than industry in Germany or Sweden. The point is that U.S. energy use cannot be "explained" away by references to structural phenomena such as those above—nor should we necessarily be ashamed of relatively high energy use in the past, given our access to relatively inexpensive fuels and electricity.

The importance of these myths in the energy debates of the past and present cannot be overemphasized. It was often argued that differences in culture, life style, policies, form of government, or nonenergy resource base affected energy use in ways beyond the control of our own energy policy—ergo, we could not imitate other countries. But international comparisons help sort out different energy technologies such as insulation techniques, industrial processes, automobile propulsion systems, etc., that have important consequences for energy use. As techniques for using energy, these systems have little to do with the nontechnical differences among peoples and countries. Certainly the implementation of specific technologies, in the name of energy conservation, *may* depend on nontechnical aspects of a country and its life style. But such questions can only be answered once energy use and conservation among many countries has been fully examined.

International Comparisons: A Few Rules

Our discussion of the myths suggests that energy analysis for the purposes of conservation is necessarily complicated—simplifications have caused too much confusion in the past. In order to make international comparisons meaningful, it is necessary, therefore, to observe certain important practices.

WHAT IS CONSERVATION?

We need to agree on what we mean by conservation. Elsewhere I have argued that energy conservation means using less expensive resources, most notably capital in place of energy, so as to reduce costs. This definition is the one which has been used by the authors of this book (see the explanation of energy efficiency in the Introduction to this volume).

To measure conservation, especially among different countries, energy use must be carefully broken down into single homogeneous activities—driving, raw steel making, metal production, etc. This breakdown is necessary if we are to measure energy intensity as energy per unit of output, a kind of inverse of efficiency for each activity. By using intensities, we avoid the need to compare "standard of living."

Usually reductions in intensity indicate conservation—less energy for a house of a given size in a given climate almost always indicates *greater* degree of weatherization in a home or less energy intensity, i.e., a greater effort toward conservation. Turning back thermostats, on the other hand, means buying less warmth, a change in the mix of consumer purchases.

Structure or variations in the kinds of uses for energy also affect energy use—these variations in the market basket of consumer purchases or in the composition of industrial output will to a greater or lesser degree reflect resource costs, of which energy's share is still relatively small for most goods and services. In the short run, changes in the cost of energy will only have slight effects on the market basket. In the long run, how-

TABLE 1. SOME BASIC DATA ON 9 INDUSTRIALIZED COUNTRIES. NOTE THAT THESE FIGURES ACCOUNT FOR CONSUMPTION BUT DO NOT MEASURE EFFICIENCY OR STRUCTURE

Consumption by Sector	*U.S.*	*Canada*	*France*	*W. Germany*	*Italy*	*Netherlands*	*U.K.*	*Sweden*	*Japan*
A. (Tons Oil Equivalent per Million Dollars GDP)									
Total energy consumption	1,480	1,772	795	1,031	915	1,272	1,121	1,062	849
Transformation losses	250	401	140	170	133	164	254	267	147
Energy sector	135	128	57	74	48	100	81	33	48
Transport sector	327	305	117	132	136	134	146	121	105
Industry sector	309	388	219	299	282	254	318	275	330
Household-commercial	374	480	223	300	220	407	271	348	164
Nonenergy use *	86	70	38	55	96	213	53	18	54
All sectors minus transport	1,153	1,467	678	899	779	1,138	976	941	744
B. (Index, United States = 100)									
Total energy consumption	100	119.7	53.7	69.7	61.8	85.9	75.7	71.8	57.4
Transformation losses	100	160.4	56.0	68.0	53.2	65.6	101.6	106.8	58.8
Energy sector	100	95.6	42.2	54.8	35.6	74.1	60.0	24.4	35.6
Transport sector	100	93.3	35.8	40.4	41.6	41.0	44.3	37.0	32.1
Industry sector	100	125.6	70.9	96.8	91.3	82.2	102.9	89.0	106.8
Household-commercial	100	128.3	59.6	80.2	58.8	108.8	72.5	93.3	43.9
Nonenergy use *	100	81.4	44.2	64.0	111.6	247.7	61.6	20.9	62.8
All sectors minus transport	100	127.2	58.8	78.0	67.6	98.7	84.7	81.6	64.5

Source: Resources for the Future. Note one ton oil equiv. = 11.63 mWh = 40.8 GJ.
* Energy raw materials used as feedstock for the petrochemical industry.

ever, changes in energy costs—and perhaps more important, changes in life style, preferences, government, or institutional policies, can affect total energy use. Labeling these changes as "conservation" is possible but requires careful examination of many nonenergy factors: are the French "energy conserving" because their total land travel is about half our own? My observations suggest that indeed some of the differences in miles traveled are due to the higher cost of fuel and autos in France relative to incomes, but more of the difference is due to urban planning and life style. Do we call this effect "conservation"? We must be very careful in interpreting structural differences that have great energy implications.

By the same token, are the Swedes less efficient than we because their most energy intensive paper and pulp sector is much (4 times) larger than ours? According to some measures (such as Commoner's in *The Poverty of Power*) the Swedes are "wasteful"—in our discussion they are not, at least not solely because of the size of the paper industry.

PRICES

In addition to understanding and measuring conservation, we must be aware of the mix and cost of fuels and electricity, remembering that some fuels may be relatively cheap to buy but expensive to use, such as coal (as Bauer and Hirshberg show in chapter 7). Less important for broad comparisons, but vital for detailed econometric or engineering comparisons, are the relative costs of other factors of production, especially capital and labor. Does the high cost of labor in the U.S. inhibit installation of double windows as a conservation measure vis-à-vis Sweden? Do Swedish industries obtain capital at such low interest rates as to make heat recovery equipment "cheaper" there than in the U.S.? A handful of studies has treated factor analysis carefully. As a first approximation, however, energy intensities and energy prices are a good measure of how effectively energy is being used in a country.

PHYSICAL COMPARISONS

Certain other determinants of a country's energy use are also important, notably climate, composition of imports and exports, age of housing and industrial stock, density of cities, and overall density of population. Indeed it is the importance of these factors that makes comparisons of total energy use among countries relatively useless—only when each factor is sorted out and quantified can meaningful analysis be performed.

POLICIES

Finally, no study of energy uses among nations is complete without at least a summary of important energy related policies. Are building codes important? Are automobiles taxed? Do standards exist for industrial techniques or are processes chosen on the basis of costs? What about environmental regulations? These policy factors may play an important role in explaining differences in energy use among countries and must be examined whenever differences in intensity suggest important possibilities for conservation.

To date, several detailed comparisons of two or more countries have appeared. While these studies differ in method and findings, a reading of all of them suggests that there is flexibility in energy use based upon international examples. To illustrate this flexibility and then illuminate some of the possibilities for conservation, we will concentrate first on some of the lessons from the other comparisons, then the U.S.-Sweden example.

Industrialized Countries

The well publicized Workshop on Alternative Energy Strategies (WAES) study attempted an integration of demand and supply forecasts from a large part of the world. As a by-product, a rather detailed set of worksheets appeared that provided the data base from which member countries' teams made their forecasts. The results are presented in rather raw form (see my review in *Technology Review*, June 1978), but carefully used they provide the first look at energy use in many countries.

Most of the countries in the WAES data base forecast increases in key consumer amenities–living space (heating) and auto use. Though conservation will be stressed in these countries as well–especially where centuries-old buildings are replaced–energy demands in total will still grow somewhat. The important exceptions are Sweden, the U.S., and Canada. There the possibility of negative energy growth in the residential sector and, in the case of the U.S., automobiles, is very real. Thus the overall energy per GNP ratios of France, Germany, Denmark, Japan, or the Netherlands will not fall as fast as those of the more energy intensive countries, Sweden, Canada, and the U.S.

The raw WAES data provide some surprising insights into international differences in energy use patterns. Americans travel far more in total than anyone else, but the predominance of the auto in every country (except Japan) is surprising, and auto use seems to increase with income, especially as a competitor to mass transit. The U.S. and Canada use considerably larger fractions of total energy for direct consumer uses (autos, homes) than nearly all other countries, though Europe and Japan are experiencing marked increases today in consumer uses. Another surprise for many observers is the degree to which industrial energy use dominates other countries' balances (first column in Table 2). Intensities of raw materials tend to be lower in Europe than in the U.S. (Table 2),

TABLE 2. ENERGY IN INDUSTRY, COMPARATIVE INTENSITIES

	Industrial Sector: Percent of Total Energy Use	Crude Efficiencies for			Steel, 1972				
					Percent of Total Energy	Furnaces, Percent			Tons/Oil Equiv./Ton Steel
		Cement	*Pulp and Paper*	*Oil Refinery*		*Open Hearths*	*Basic Oxygen*	*Electric*	
Austria	30.7	21.5	14.4 *	—	—	—	—	—	18.8
Canada	26.0	22.7/35.4	28.2	—	2.5	39.5	43.9	16.6	16.3
Denmark	16.8	39.2	14.8 *	12.7	—	—	—	—	—
France					7.5	43.5	45.9	10.6	22.2
Germany	36.1	21.7	18.3	19.1	9.6	25.2	64.6	10.2	23.4
Italy	34.4	23.0	14.2 *	10.7	6.1	20.1	39.1	40.8	16.7
Japan	47.3	28.6	21.4	11.0	14.9	2.0	79.4	18.6	19.7
Netherlands	32.5	31.4	—	—	4.4	4.2	88.9	6.9	20.5
Norway	33.9	27.5	27.6	—	—	—	—	—	—
Sweden	31.6	33.4/20.1	20.5	11.7	5.6	21.3	36.5	42.2	19.3
Switzerland	18.3	—	23.3	—	—	—	—	—	—
United States	27.7	38.6	24.2	21.4	3.9	26.2	56.0	17.8	23.9
United Kingdom	28.5	33.1	26.2	17.5	7.3	37.9	42.7	19.4	25.5

All quantities in GJ/metric ton. Steel data from RFF, other data (incl. Austria) from IEA. No corrections for product mix or inputs, so figures should be taken as illustrative only. Where two figures are given for cement, the former is for *wet,* the latter for *dry* processes.

* Paper so marked probably excludes some energy embodied in purchased pulp, and Canadian/Norwegian figures are affected by electricity accounting.

but output of these materials is somewhat greater in Europe, erasing some of the potential differences in *total* energy use. None of these differences arose out of any "moral war" to save or "waste" energy—autos and gasoline have been taxed elsewhere for fiscal reasons, while industries in Europe, being newer than our own and built at a time of considerably higher energy prices, simply combined resources rationally, using less energy to produce important energy intensive materials.

A study titled *How Industrial Societies Use Energy* was able to offer more analysis of data and policies in asking why nine countries with somewhat comparable standards of living used such varying amounts of energy. This study, carried out at Resources for the Future (RFF), shed more light on the fascinating tale of energy and Gross National Product. In working with the RFF authors, I discovered that Sweden was in many respects an unusual energy user because of energy intensive industries, climate, proliferation of autos beyond most other countries in Europe, and dwelling space. The RFF group is at present updating all their data to follow the U.S., West Germany, and Sweden since their base year of 1972. Nevertheless, the important conclusions of their first book (embellished by WAES and other data) should be summarized here. (See Table 3.)

1. Transportation differences account for nearly one half of the variation in the energy/GNP ratio among countries, with the remainder divided up among residential/commercial energy industries and transportation. *Differences* in total energy use in manufacturing were small because greater efficiencies in Europe are offset somewhat by greater output of raw materials compared to the U.S. (See Tables 1 and 3.)

2. American households spend a larger fraction of their incomes on direct energy purchases than do European, at least before the embargo. Not surprisingly, U.S. prices were the lowest. (Some prices are shown in Table 4.)

3. After adjusting for climate, the U.S. uses 40 percent more energy for space comfort than the average in Europe. Part of the reason is the larger, detached homes in the U.S. vis-à-vis the Continent. Moreover, full house heating is only saturated in North America and Scandinavia. Efficiencies were significantly better in Sweden, with Canada lying between Sweden and the U.S.

4. In the transportation area (Table 5), total travel, efficiency of each mode, and (to a lesser extent) the actual mix of modes vary widely. The auto dominates everywhere (except in Japan), but higher urban densities, higher auto costs, higher fuel costs, and other measures have kept the absolute level and relative share of heavily subsidied public transit at two to six times the U.S. level per capita. In the freight sector, higher U.S. consumption is explained mostly by larger volumes of freight moved farther, because the U.S. system appears to be more reliant on rail over-

TABLE 3. A SCHEMATIC RANKING OF FACTORS AFFECTING COMPARATIVE ENERGY CONSUMPTION/GDP RATIOS, BY COUNTRY, 1972 (FOR A GIVEN CATEGORY, THE LOWER THE NUMBER, THE GREATER THE EFFECT ON RAISING THE ENERGY/GDP RATIO)

Factors	*United States*	*Canada*	*France*	*West Germany*	*Italy*	*Nether-lands*	*United Kingdom*	*Sweden*	*Japan*
Energy prices (lowest prices = 1)	1	2	4	5	9	6	8	3	7
Passenger miles per unit GDP	1	5	9	3	2	6	4	6	8
Percentage of passenger miles accounted for by cars	1	2	8	4	7	5	6	3	9
Energy consumption per car-passenger mile	2	1	4	8	5	7	9	6	3
Cold climate *	7	2	5	4	9	3	5	1	8
Size of house & percentage single family	1	1	6	6	8	5	4	3	8
Extractive industry GDP as percent of total GDP	2	1		4	6		3	5	7
Industrial GDP as percent of total GDP	7	8	2	1	6	4	5	9	3
Ratio of industrial energy consumption to industrial GDP	2	1	9	8	6	5	4	3	7
Degree of energy self-sufficiency	2	1	7	5	8	3	4	6	9
For reference:									
Energy/GDP ratio	2	1	9	6	7	3	4	5	8
Energy per capita	2	1	7	5	9	4	6	3	8
GDP per capita	1	3	4	5	9	6	8	2	7

* Climate measured by degree days. Source of ranking, RFF.

TABLE 4. SOME IMPORTANT ENERGY PRICES

	Gasoline ¢/Liter, 1977		*Low Sulphur Heating Oil $/GJ*		*Dom. Electricity ¢/kWh, 10,000 kWh/yr*	*Dom. Gas $/GJ*	
	Without Taxes	*With Taxes*	*1972*	*1977*	*1976*	*1972*	*1976*
Austria	19	38		4.67	6.0		4.44
Belgium	16	37		3.52	5.1		5.00
Canada			.96	2.22	1.8	0.91	1.39
Denmark	17	41		3.64	4.5	—	5.83
England	12	25	1.41	2.69	2.7	2.89	2.75
France	16	40	1.22	3.58	3.5	3.30	6.11
Holland	17	39	1.03	3.47	4.4	1.70	3.05
W. Germany	15	35	0.89	3.19	3.8	3.09	5.28
Italy	14	51	1.03	3.06	5.3	1.91	2.22
Japan			1.15	3.33	5.6	—	6.90
Norway	16	39		3.72	2.5	—	—
Sweden	17	35	1.24	3.44	2.3	—	—
Switzerland	15	36		3.08	2.7		6.94
United States	13	16	.74	2.78	4.1	1.12	2.50
(For 1972, see Table 5)							

Approximate prices in US¢/liter: *Sources of Data,* Industriverk, Stockholm, RFF, OECD. 4.7 Swedish crown = 1$, 1977.

all. This reliance on rail is due to structural or historical or geographical or political reasons as explained in the Hemphill chapter on transportation, and it appears (to me) to have little to do with energy costs.

5. In industry, it takes more energy in the U.S. to heat a given material than in most countries in Europe. (This confirms the WAES implications.) Obviously, other factors, such as the cost of capital and labor (see Hogan's analysis in Chapter 1) influence the choice of technologies used in industries. Table 2 includes the RFF data for the steel industry.

My own view of the RFF evidence, and many other studies, suggests that higher fuel prices in Europe have made more energy efficient technologies profitable—elsewhere the evidence points to these technologies, especially in steelmaking and cement, being more productive with respect to all resources, not just energy. Thus the energy conservation characteristics of European heavy industry do not appear to arise simply because more labor is used. Instead, capital may substitute for energy. Clearly the RFF study, combined with other evidence, offers proof of the existence of many energy using technologies and processes in industries

TABLE 5. PASSENGER TRANSPORTATION: 1972

	Pass-MI/Cap	MI/Auto	Percent Auto	Energy/Cap MwH Cap	Intensity	Gas Price (US=100)	Percent of Income	Auto Ownership, Cars per 1000 People	
								1961	1972
United States	11,300	9,360	92	9.4	.90	100	3.4	344	462
Sweden	6,280	8,900	84	(3.8)	(.60)	(180)	(0.8)	173	303
Canada	6,550	10,000	88	6.3	1.1	(110)	—	237	377
France	3,980	—	77	2.2	.71	256	0.7	133	269
W. Germany	5,870	8,900	82	2.4	.51	243	1.1	92	253
Italy	4,160	7,610	80	2.2	.65	348	0.6	48	229
Netherlands	4,620	10,000	81	2.2	.59	—	—	53	229
Japan	4,990	8,950	80	2.0	.49	192	1.1	113	230
United Kingdom	3,760	—	34	0.9	.74	250	0.2	7	119
Europe avg.	4,840		80	2.3	.60	—	—		

Source: RFF; IEA; Swedish data modified by Schipper and Lichtenberg; prices for gasoline, income shares from RFF; distance/auto/yr from WAES.

Passenger transportation: Shown are the total miles, the share taken by autos, the resulting per capita energy consumption, the intensity in kWh/passenger mile, the gasoline price relative to the U.S., and the percentage of income spent on driving. Finally, auto ownership figures for 1961 and 1972 are shown, displaying the rapid growth in Europe and Japan that still lies far from saturation.

that would save energy when employed in the U.S. Further investigations will reveal which technologies are attractive.

In short, the RFF study finds much variation in energy intensities among important tasks. Not all of this variance embodies conservation, but the *flexibility* of energy use is clear.

The U.S.–Sweden Comparison

Neither the WAES study nor the RFF survey fully breaks down energy use in any of the countries surveyed. At the time these two surveys were underway, however, I began (with my colleague Allan J. Lichtenberg) comparison of energy use in Sweden and the U.S. This work has been continued by others, leading recently to a comparison of energy use in individual towns in Minnesota and Central Sweden. These energy profiles reveal many details not available to the wide area surveys. Though many are weary of the "Swedish example," Sweden provides an interesting comparison with the U.S. precisely because so many features there are close to those in the U.S. Sweden also exhibits certain extremes, like climate, but lies median to other European countries in other areas. Because data on energy use in Sweden are excellent, many ideas could be reviewed that only arose more recently in studies of other countries. It is, therefore, worth reviewing the highlights of the comparison. (Readers wishing greater detail should turn to *Science*, December 3, 1976.)

To compare any two countries, many small effects have to be accounted for before energy use could be directly compared. These effects include differences in natural distances, fuel extraction (almost nonexistent in Sweden), and climate. (Air conditioning is nonexistent in Sweden, but there is little need to heat factories in the U.S., and these two uses, by coincidence, nearly compensate.)

An additional consideration, often overlooked, turns out to be important. If one counts the energy embodied in the goods and services making up foreign trade, it is found that the U.S. is a slight importer of energy, in an amount equivalent to 1 percent of the total energy use in 1973. This includes the energy used to refine fuels that are imported and exported, but not the thermal energy of combustion contained in those fuels. Sweden, in contrast, is clearly a net exporter of embodied energy, with the net embodied energy amounting to 8 to 9 percent of total internal consumption. This is also true for West Germany and Japan. On the fuel side, Sweden, like most European countries, imports a larger share of her energy, both crude and refined, while the U.S. imports considerably less in relative absolute terms per capita. The U.S. exports coal and Sweden exports refined oil because of excess refining capacity. Moreover, geography and trade put certain uses of energy out of reach of the normal accounting practices since a much larger share of Swedish production, consumption, and travel passes through foreign countries than is the case for the U.S. Fortunately, the most troublesome discrepancies or difficulties turn out to be relatively small or readily quantifiable.

After allowing for these adjustments, it is found that the greatest differences in energy use appear in the intensities (or efficiencies) of use for process heating, space heating, and transportation. The overall effects of both intensity and mix of output (for Sweden and the U.S.) are displayed in Table 6. As can be seen, space heating in Sweden is remarkably less intensive than in the U.S. when measured in Btu per square meter per degree-day. Other studies suggest that Sweden and Denmark are unique in this area. The living space per capita is nearly as large in Sweden as in the U.S., while most of Europe falls behind these countries in this important measure of living standards. The energy intensity of apartment heating in Sweden is nearly as great as that in single-family dwellings (see below). This means that the relative efficiency of space heating in Sweden vis-à-vis the U.S. *cannot* be ascribed to the greater proportion of apartments there compared with the U.S.

On the other hand, households in Sweden generally have fewer appliances than in the U.S., reflecting a different life style and lower after-tax incomes, and this results in a lower household use of electricity. Other European countries fall even farther away, though the gap is narrowing. In the commercial sector, the same high degree of thermal integrity appears in Sweden.

Indoor temperatures in Sweden are higher than in the U.S. One relative inefficiency in the use of heating and hot water occurs in Sweden because of common metering and unregulated hot water and heating systems. This leads to a surprisingly large consumption of fuels for heating in apartments, although the overall use of heating is more efficient in Sweden than in the U.S. because building shells are well constructed.

In the industrial sector, the differences in intensity are consistent with the results of other studies. Sweden is neither the most nor the least efficient. While oil refineries in Sweden produce relatively less gasoline than in the U.S., other product mixes are comparable. The overall Swedish mix in manufacturing is weighted more heavily toward energy intensive products than is the case in the U.S. The lower energy intensities found in Sweden, however, are generally tied to higher energy prices there, suggesting that prices do affect industrial energy "needs" considerably.

The greatest contrast is found in transportation, dominated in both countries by the auto. Swedes travel only 60 percent as much as Americans and use 60 percent as much fuel per passenger mile. This held Swedish gasoline use in the early 1970s to one-third of our own. Mass transit and intercity rail are less energy intensive and more widely used in Sweden, while air travel is overwhelmingly larger in the U.S. Intracity trucking in Sweden is considerably less energy intensive than in the U.S.,

Table 6. Sweden/U.S. Contrasts in Energy Use; Ratios Are Listed

		Per Capita Demand	*Intensity*	*Total Energy Use*	*Notes*
Autos		0.6	0.6	0.36	Swedish 24 M.P.G. driving cycle uses less energy
Mass transit trains, bus		2.9	0.80	2.35	Mass transit takes 40% of passenger miles in trips under 20 km in Sweden
Urban truck		0.95	0.3	0.28	Swedish trucks smaller, more diesels
Residential space heat (energy/deg. day × area)		(1.7 ×0.95)	0.5	0.81	Sweden 9200 deg. days vs 5500 U.S. deg. days
Appliances		?	?	0.55	U.S. more, larger appliances
Commercial total/sq. ft.		1.3	0.6	0.78	Air conditioning important in U.S. only
Heavy industry (physical basis)	Paper	4.2			Sweden more electric intensive due to cheap hydroelectric power. Also Swedish cogeneration
	Steel	1.1			
	Oil	0.5	0.6-0.9	0.92	
	Cement	1.35			
	Aluminum	0.5			
	Chemicals	0.6			
Light industry ($V.A.)		0.67	0.6	0.4	Space heating significant in Sweden
Thermal generation of electricity		0.3	0.75	0.23	Swedish large hydroelectric, cogeneration

Listed are the ratios of Swedish/U.S. demand, intensity, and energy use totals. Differences in demand account for structural differences (market basket). Differences in intensity tend to indicate conservation. For transportation the demands are measured in passenger or ton miles.

but long haul trucks in Sweden use slightly *more* energy per ton-mile than in the U.S. The greater distances in the U.S. mean that ton-mileages (at distances greater than thirty miles) are far greater there. The overall U.S. long haul mix is less energy intensive, but total use is greater because of distance. Here is a clear example of how greater *use*, on the part of the U.S., has little to do with inefficiency. In fact, the American freight machine is more weighted to less energy intensive railroads than in most other countries.

Historically, higher energy prices in Sweden than in the U.S. are an important factor that has led to the more efficient energy use in that country. While preembargo oil prices in both the U.S. and Sweden were roughly equal (Table 7), Americans enjoyed natural gas and coal resources that provide heat at a 20 to 50 percent lower cost compared to oil. In the case of electricity, the two countries were radically different

TABLE 7. TYPICAL ENERGY PRICES IN THE U.S. AND SWEDEN. EXCHANGE RATE USED IS $1 = 5.18 SKR (1960–1970) AND 4.30 SKR (1974)

	U.S.[a]			¢/kWh	*Sweden*[b]			¢/kWh
	1960	*1970*	*1974*	1970	*1960*	*1970*	*1974*	(1970)
Oil products (¢/gal)								
Gasoline[c]	30	35	45	*1.04*	53	61	116	*1.82*
Diesel	23	28	35	*0.83*	42	48.8	90	*1.45*
Heating oil								
Small customers	15	18	35	*0.50*	13.3	13.2	40.6	*0.37*
Large customers	10.5	12	25	*0.33*				
Heavy oil	7	8	23	*0.23*	7	8.5	22.5	*0.24*
Gas (¢/MM Btu)								
Residential	82	87	113	*0.29*	—	550	680 *	*1.9*
Industrial								
Firm service	51	50	—	*0.17*	—	—		
Interruptable service	33	34	—	*0.11*	—	—		
Coal, industrial[d] ($/ton):	10	13	25	*0.14*	—	18		*0.2*
Electricity (¢/kWh)								
Base	2.75	2.75	—	*2.75*	3.14	2.12	2.3†	—
Base and space heating	1.75	2.0	—	*1.5*	—	—1.5	2.0†	
Industrial	1	1	1.5	*(0.4-2.1)*	—	0.93	1.8†	*(0.6-2.2)*

[a,b,c,d] Data sources listed in Schipper and Lichtenberg.
* For 1973.
† For 1975.

(up to 1972). Since 75 percent of all electricity generated in Sweden was produced by hydropower, the ratio of the cost of electricity to the cost of heat from fuel was only *half* as great in Sweden as in the U.S. Industry in Sweden naturally developed a more electric-intensive technology base. However, 30 percent of thermal electricity generation in Sweden was accomplished through combined production of useful heat and electricity in industries or in communities, the latter systems providing district heat. Consequently, in Sweden only about 7,000 Btu's of fuel were required in 1971–72 for the thermal generation of a kilowatt-hour of electricity. Increases in the cost of nuclear electricity and oil favor the continued expansion of combined generation, but institutional problems have slowed that expansion in the late 1970s.

An example of the effect of different prices helps explain Swedish energy use. In Sweden, autos are taxed in proportion to weight, both as new cars and through yearly registration. Swedes found a loophole, the registration of autos through companies; but the government discovered this trick and raised the tax on company owned cars. Gasoline is taxed, the amount recently being raised to ninety cents per U.S. gallon versus less than fifteen cents in most of the United States. Even still, Sweden has a relatively low priced gasoline market compared with France or Italy (Table 5). But overall high prices, compared to the U.S., restrain total auto use, especially in short trips and in cities.

Although the impression that Sweden is somehow "energy wise" and the U.S. less so is unavoidable, the real lesson from this two country comparison is that energy use for important tasks is flexible, given time, technology, economic stimulus, and, in some cases, favorable government or institutional policies. Indeed Sweden could be using more energy than the U.S. per capita (or per unit of GNP) and still be more efficient (as is the case in manufacturing) or the converse. And Sweden is not a special case. Other countries show similar variations in energy use indicating that the Swedish example tends to reinforce the findings of the broad survey.

Other Countries since the Embargo

What has happend since 1973? While in many respects it is too soon to tell, the evidence suggests that countries are conserving energy. At least in the postembargo period, the amount of energy required to produce a unit of GNP growth has declined.

While this result holds for nearly every wealthy country, it is still subject to all of the vagaries of E/G statistics. For example, one International Energy Agency (IEA) study showed that during recessions the U.S. E/G tended to rise, while E/G in most other countries fell. One explanation is that the E/G in the U.S. is dominated by consumer uses of energy that only fall in extreme recessions, while in Europe E/G, dominated by commercial and industrial uses, contracts more quickly with economic slowdowns. In Sweden, for example, 1974 was a boom year for the economy, but real conservation efforts resulted in a 9 percent increase in industrial production with only a 1 percent increase in energy use. In 1976, however, industrial use fell due to recession. In the United Kingdom energy use has fallen absolutely while GNP has increased. Clearly, these contrasting events must be separated in any discussion of conservation.

Nevertheless, the IEA data do suggest that people, firms, and institutions are using energy more efficiently. Table 8, while relying on the energy/GNP ratio as an indicator, does give the impression that energy use and economic activity have a different relationship now than previously. The marginal increase in energy consumption needed to raise Gross Domestic Product (GDP) by $1,000 is shown for two periods, and the most recent five years show a clear decline in all but one country—Spain. Most important, these changes, signifying at least some significant conservation, have taken place without an overwhelming change in capital stock, though the marginal nature of the statistic applies to new activity. As existing stock is replaced or retrofit, the overall E/G should fall even more than it has so far. While a sector by sector analysis has yet to be done for most countries, data from Sweden and Denmark indi-

TABLE 8. GROSS ENERGY PERFORMANCE

	Energy and Growth	
	1960–1970	*1972–1977*
IEA	1.05	0.48
W. Germany	0.99	0.30
Japan	1.02	0.67
Italy	1.59	0.31
Holland	1.65	0.43
Spain	1.10	1.63
Sweden	1.08	0.81
United Kingdom	0.73	negative
United States	1.11	0.34
France	0.86	0.54

Conservation Performance: Tons of oil equivalent required to increase GDP by $1,000 during pre and postembargo periods. Data from the IEA. Note that the coefficient for the United Kingdom in the second period is undefined because energy use contracted while GDP increased.

cate that economies in the housing sector have been significant. In the latter country, 19 percent of all roofs were insulated in 1974 and 1975! All countries have shown improvements in industries, and the U.S. has shown a dramatic improvement in the energy consumption of passenger autos. Conservation, even before large-scale campaigns involving investment have been mounted, appears to have been anchored in the economies of the largest or richest industrial countries.

At the same time, structural changes are still reshaping the patterns of energy use toward somewhat more energy intensive activities. Autos are still growing in importance in a country like Germany, where auto ownership was relatively high and freeways extremely well developed even before the oil embargo. There may be a subtle reason why the auto competes so well with one of the best and fastest rail systems in the world—speed limits are unknown on the German autobahns. Speed limits in Germany have proven politically impossible, and the net result is that automobiles can travel point-to-point faster than even the fastest trains in many cases, since much transit time (home-to-station, stopovers) is eliminated. In Sweden, by contrast, several 200 to 400 kilometer stretches that are well developed from the motorist's point of view are nevertheless faster by train because Sweden has a speed limit of 110 kilometers per hour on freeways and 90 on other main roads. Thus the rail corridors Malmö-Gothenberg, Malmö-Stockholm, Stockholm-Gothenberg, and Stockholm-Luleå still compete with autos for long distance travel.

Of course, a 200-plus kilometer per hour train system could be built in Germany (as has been working in Japan), but then the energy saving qualities might be reduced or eliminated. Germans are not necessarily "wasting" energy by abandoning the bundesbahn for the autobahn because they are gaining time and convenience. This subtle example, then, illustrates well the problems facing the so-called "efficient" nations.

What Can We Learn from Other Countries?

While much of the value of studying the energy habits of other countries lies in the understanding we get about how energy economies work *in theory,* it is useful to review practices in foreign countries that may be of practical help for American energy users. It should be emphasized, however, that we must work to find our own solutions to energy uses that are climate dependent (buildings), life style dependent (land use planning and transportation), or policy dependent. Even in industry, it is important not to simply shop in other capitals for energy saving equipment, but to build instead on existing ideas with new ideas, leapfrogging existing efficient technology to find even more productive ways of using energy with other resources.

BUILDINGS

Only the Scandinavians widely display truly enviable practices in the buildings sector There, energy use per unit of area per unit of climate, the best measure of efficiency, is truly less than in Central Europe or the United States. Moreover, evidence accumulated in conversations with European energy officials suggests that most European countries are a long way from establishing truly energy efficient building stocks.

Scandinavian home building practices make it clear that we can reduce building energy needs considerably—by as much as 80 percent compared with pre-1973 homes in the United States. While insulation of existing homes in the U.S. is the most popularly cited need, control of infiltration and ventilation may make an even larger contribution to saving energy profitably, when existing or new Swedish buildings are compared with untight U.S. structures. Experience in the building research programs at the Center for Environmental Studies at Princeton and the Lawrence Berkeley Lab suggests that we can achieve the low air infiltration rates now called for in Swedish or Danish building codes (considerably less than one air change per hour in homes).

One effect of careful insulation and tightening of structures is the increase in comfort that goes beyond the relief of a lower heating bill. When structures are carefully controlled the heat comes on less, causing less air exchange and heating up of the indoors near vents. Drafts are reduced. The temperature difference between floor and ceiling and between areas near windows and inner parts of rooms is reduced, reducing both air motion and discomfort. Indeed there is the suggestion that Swedish homes are built well so as to satisfy desires for comfort ahead of simply saving energy.

One of the most important technologies being applied in Scandinavia is the heat exchanger. It has become clear that infiltration losses in homes can be reduced so far that odors, indoor pollution including evaporated plastics, radon gas from building materials, and cigarette smoke can become a nuisance or even a true health hazard. Forcing ventilation by fans and ducts has been a common practice in Swedish homes: the exhaust air contains valuable heat, however, and an inexpensive heat exchanger could recover much of the heat while allowing the unpleasant pollutants to be exhausted before they could build up in the home. In new Swedish apartment buildings, where the heat content of exhaust air is enormous, heat exchangers can be required, an attractive possibility for centrally heated and ventilated U.S. buildings. It should be noted that heat recovery is extremely important in the U.S. in warm months, when coolant from exhaust air can be recovered in the system. At present, several European and Japanese firms are planning to market inexpensive heat exchangers in the U.S.; undoubtedly U.S. manufacturers will be close behind, and soon ahead.

INDUSTRY

Why do European process industries use less energy/output than our own? For one thing, European capital is newer, since economic growth in the postwar period was greater than in the U.S. But newer equipment in both Europe and the U.S. uses less energy per unit of output than older equipment. Thus from a structural point of view, Europe has an "advantage." The price of energy has been a big factor. Since energy is an important cost factor in the paper, cement, steel, and chemical industries, their areas show (not surprisingly) the greatest energy differences, particularly the paper industry. However, it is important to add that both the U.S. and Europe are improving since the embargo because the industrial energy and power prices in both countries have risen considerably. Thus, a new large dry oven installed in Slite, Gotland, Sweden, undercuts all existing Swedish cement ovens in terms of energy performance. But the same is true of new American facilities, according to a report in the *New York Times* (January 1, 1978). Thus we are learning.

One topic of controversy that has attracted considerable attention because of multicountry comparisons is the role of factor substitution in industrial energy use. Several studies have produced apparently conflicting results as to the possible substitutability of capital for energy. When specific energy using industries are studied, however, the energy saving role of advanced technologies that substitute capital for energy appears clear. This is particularly true of process industries where heat, not electricity, dominates energy use. Thus the dry cement oven, which actually costs less (in investment/output) replaces the wet oven in most countries now, allowing increases in productivity and in energy efficiency. In Sweden, at present, an intensive program of investment in energy saving devices in existing industrial plants is underway. Most countries show a marked decline in energy/output for important materials over time (see Figure 1).

A recent comparison of steelmaking in the U.S. and Japan showed clearly through both process analysis and economic factor analysis why Japanese steel mills use less energy per ton of output. This investigation noted differences in the steelmaking process—the U.S. still uses considerable numbers of energy (and pollution) intensive open hearth furnaces while the Japanese rely on a considerably higher fraction of basic oxygen furnaces. The latter produce more steel per man hour and per unit of energy. Less Japanese steel was "wasted" in each process or lost so that overall output was also greater in a Japanese mill. In the final stages of forming, the continuous casting process, by which hot steel is formed

Fig. 1. Energy Consumption for Cement

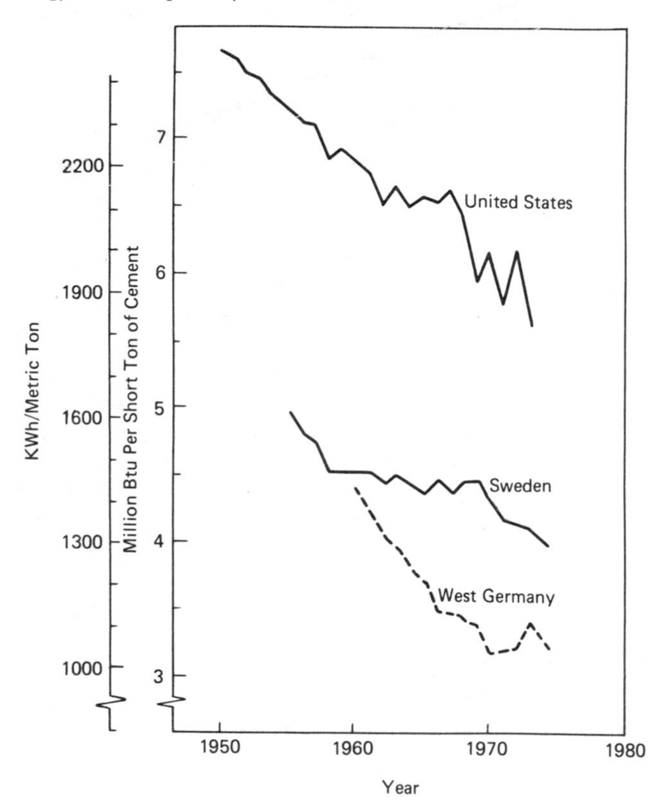

immediately into intermediate or final products rather than being cooled down (and later reheated with natural gas) was also important in accounting for the 25 percent energy saving in Japan. As reported in *Energy*, August, 1978, T. V. Long and his colleagues found that overall Japanese steelmaking employed less labor and slightly more capital per unit of output than American steelmaking. The Swedish example lies intermediate to these two.

Long and his colleagues looked at steel and papermaking for a number of countries and found again that capital can substitute for energy in most industrial countries. The same workers also measured the relationship between energy and capital across time and for all manufacturing in most European countries and found again that capital was a clear substitute for energy. They noted that this held both in the high priced energy countries like Japan and West Germany and in the U.S. Thus international comparisons point clearly to opportunities for the U.S. to look at a variety of industrial technologies that are profitable when energy prices climb.

The clear lesson from foreign experience is that technology, prices, and the actual composition of natural resources available to a country share in determining the energy use and conservation characteristics of industry. Other countries pay dearly for fuels, so these resources were "conserved" relative to the United States. Historically, conservation also accompanied gains in productivity.

One important topic being discussed in Europe and Japan is structural change. Should governments encourage or discourage the growth or decline in the most energy intensive industries, ones that would be most vulnerable to energy supply interruptions in the future? The question is particularly important in the European steel community and in the Nordic paper industries. The U.S. faces similar questions in the steel and aluminum industries, especially as sources of cheap electricity literally dry up for the latter. While such questions are beyond the scope of the present discussion, they must be kept in mind because they can have great influence on total energy demand in the long run, as pointed out in Chapter 1 of this book.

Cogeneration of electricity and heat in industrial plants is receiving increasing attention in the U.S. In this area Sweden leads most of the world, particularly because of the size of the paper industry. However, even in Sweden "cheap" electricity has hindered the expansion of cogeneration, especially just now when long-term electricity contracts hide the marginal cost of new power from existing plants. About 5 percent of Sweden's 1972 electric power production came from industrial cogeneration, 4 percent in 1976 (a recession); and estimates of 8 to 10 percent have been made in studies of the 1985 period and beyond for the United States. For Germany, France, and Italy total cogeneration in industry amounts to around 15 percent of total electricity production, according to recent Organization for Economic Cooperation and Development (OECD) figures.

However, cogeneration in Sweden has always been regarded as a method of using waste heat from electric power production. A much larger potential that is untapped in both Sweden and the U.S. arises when one makes electricity whenever high temperature (or low temperature) heat goes unused, the electricity being sold to the utility net whenever it is not needed on-site. American estimates find a potential of 5 to 15 percent of all electricity in 1985 can be generated this way with good profitability for the generator. Institutional factors may inhibit this growth in an energy conserving technology, but the foremost barrier is still the unrealistically low price of electricity to industrial users. Additionally, the uncertainties over environmental regulations, lower prices for natural gas and oil than in Sweden, and fear of utility/industry cooperation figure in the muzzling of the cogeneration potential in the U.S. International comparisons of experience in cogeneration deserve more study now to determine which existing arrangements are most attractive for the future. This applies both to technical and to institutional problems.

DISTRICT HEATING

One technology often suggested by the European experience is district heating (DH), by which blocks (or square kilometers) are provided with water-borne heat (and hot water) from central plants. In Denmark, Sweden, Berlin, and many eastern European countries a significant fraction of all apartments and most buildings in city centers are heated by district heating. Elsewhere, district heating makes important local contributions, though in no case do the energy savings attributable to DH make more than a small impact on total energy consumption.

How does district heating save energy? Heat-only systems produce hot water in well maintained high temperature boilers whose heat transfer from fuel to water is significantly higher than in individual boilers, more than offsetting the relatively small (equal or less than 10 percent) losses in transmission of water. In the ideal case, the largest possible fraction of hot water is made in conjunction with electric power. Heat that would have been rejected to the environment is now used to heat buildings; the extra amount of energy added to this water (or alternatively the electric power sacrificed) is typically five to eight times less than the useful heat produced. Alternatively DH can be described as a system that produces electricity for far smaller losses than in condensing-only power plants. Energy savings equal the extra fuel required were electricity and heat made separately. Exactly how large a fraction of all district heat is produced with electricity varies, depending on the characteristics of the heating season (or need for cooling), as well as the electric power demand characteristics and existing power plant mix. DH economics depend both on this accounting and critically on the capital cost of distribution, which in turn is very dependent upon the amount of heat sold per square kilometer. In dense areas with long heating seasons, such as cities in Scandinavia, DH provides low cost heat.

Other important advantages accrue to cities with DH. Pollution from burning oil is clearly reduced because controls are better than in separate boilers. This advantage was important in starting up such systems in Sweden in the days when oil was cheaper. Moreover, oil-fired DH systems run on cheap heavy oil. Additionally, DH centrals can run on a variety of fuels, including wood or coal, and can be built to switch rapidly. Since the combustion operation is centralized, congestion associated with delivery of fuel is minimized. Finally, DH relieves individual building owners or occupants from worrying about heating, and reliability is good.

Whether DH is a good buy for the U.S., however, is questionable. Winters here are shorter, though often colder during peak times, than in Scandinavia. When comparisons are made of DH economics in Europe and the U.S., the heating load that enters in the U.S. calculation is often assumed at today's levels, rather than calculated based upon conservation that would be appropriate at the price charged for DH. But European figures for heat demand are bloated by the lack of individual meters, a problem particularly acute in Sweden. The real cost of DH may be unknown since the unit price is so sensitive to the number of units over which the enormous fixed costs are spread. If DH can provide cooling, of course, the economics change considerably since such cooling reduces electric peak loads and reduces waste heat loading in the summer in cities. Certainly technical studies and actual implementation, as have been discussed for cities in Minnesota and other colder states, are important. At present, it appears that it is far cheaper to save fuel by end use reduction than by DH.

However, the real problems for DH in the U.S. may be institutional. Sweden and Germany have contemplated mandatory hookup laws as a means of insuring high density and thereby lowest costs. Land use planning with long time horizons, far more prevalent and accepted in Europe, is essential to the orderly build-up of a system over a decade. Moreover, DH has penetrated principally apartment areas. In Västerås, Sweden, where virtually all single and multiple family dwellings receive district heat, unit costs for detached houses were two to four times greater than apartments, both because of higher distribution costs and considerably greater losses per dwelling. In the U.S., detached homes dominate, and little high density new construction is on the horizon. DH may not fit into our living patterns except in existing downtown areas, possibly with urban renewal.

Will European DH systems be important in the U.S.? Unfortunately many of the advantages appear only indirectly and not as direct cost reductions, especially when conservation reduces heat needs so much in most of the U.S. And DH can only appear as a result of coordinated action, with government present at nearly every stage. Indeed it has been argued that DH has been attractive in many places precisely as an extension of municipal power into the service of comfort. But haggling over every recent government effort does not speak well for DH.

Thus, DH faces institutional tangles that may only be worth overcoming in areas like Minnesota, where the potential benefits are inarguably great. Smaller ventures, such as time of day pricing and individual metering of apartments or large-scale retrofit insulation programs ought to be tried first before any large-scale DH is promoted on a national scale, as is often advocated here and in Europe. For ultimately, the energy saved per unit investment should be far higher with simpler schemes than district heating.

TRANSPORTATION

In transportation the lessons for the U.S. are ones of a sensitive policy nature. The difficulty in dealing with transportation as incomes rise and autos become more important is clear: autos are popular. Obviously one cannot "hold back" the auto in the U.S. without offering attractive alternatives.

Because of low gasoline prices, tax subsidies for owning single-family dwellings, little or no land use planning, and easy access to freeways, mass transit in America seems hard put to capture any more than a small fraction of land passenger miles. The decline of mass transit's share of passenger miles in Europe, very much similar to what was seen in the U.S. twenty to forty years ago, emphasizes this point even more clearly. As usual, this decline in the mass transit *share* of traffic happens because the auto increases its absolute role in traffic. New owners, new patterns of commuting, and new uses of the auto for vacations have become as abundant in Europe as in America in the postwar era. Thus auto miles have increased tenfold in Sweden since 1950, and similar increases have occurred everywhere in Europe (see Table 5).

Clearly, strategies to save energy in this area should concentrate first on efficient, light autos. The miles per gallon standards in effect in the U.S. and Canada further this end, though, in my view, taxes on weight ("gas guzzler taxes" and gasoline taxes) would be helpful in accelerating progress beyond the nominal goal of the miles per gallon standards. The latter is particularly important during the present stagnation of gasoline prices—efficient vehicles lower the out-of-pocket cost of driving, making mass transit or walking less attractive, especially on short trips. In Europe, where auto size has increased somewhat, the spread of manufacturing of autos among several countries and strong resistance within countries make standards on miles per gallon unlikely, but higher gasoline prices have already brought on new light vehicles like the Volkswagen Rabbit, Renault's "Le Car," or Volvo 343. Nevertheless, there is still much room for improvement in the European vehicle fleet.

Technologically, however, European and Japanese manufacturers offer autos with high miles per gallon (Rabbit, Honda) that have stimulated American manufacturers, airplanes (the Airbus), and light diesel trucks. A recent ad in *Business Week* appealed to the American shipping market by pointing out that there were far more energy thrifty light diesel trucks in Germany than in the U.S.A. The U.S.-Sweden comparison suggested the consequences—short-range intracity trucking consumes far less energy per mile in Europe than in the U.S. Here is an aréa ripe for technological development.

Clearing cities of as much auto traffic as possible, leaving buses, taxis, and light trucks, would also improve freight handling and energy use, reduce pollution and congestion, and allow more efficient use of autos that remain. Such a practice is very attractive in the old capitals of Europe, especially Stockholm, but meets with jeers in the U.S. Stockholm has considered city-gate tolls, but Berkeley, California, challenged a system of traffic rating barriers twice in municipal elections, though the barriers won each time. The lesson is clear—transportation policies that affect patterns of transportation must be introduced carefully and slowly, again with an eye to the very long run, lest local opposition from consumers and businesses crush even the best of intentions. These could have enormous energy payoffs by decreasing the need to travel.

CONSERVATION POLICIES AND PROGRAMS

What about conservation policies in IEA member countries? All countries profess an interest in conservation, but only some have actively supported conservation through the marketplace, through financial arrangements, or through effective standards. Thus the IEA survey, "IEA Reviews of National Energy Programs" (June 1978), shows that while the U.S. still "enjoys" the lowest energy prices, most countries still struggle with less than marginal costs for one form of energy or another. Some countries like Germany, Switzerland, or the U.S. must await cooperation from regional jurisdictions before certain kinds of legislation, such as building codes, can pass. Sweden, Denmark, and Holland have begun campaigns of loans to industries wishing to modernize energy use, while Sweden, Denmark, Holland, France, and West Germany make loans and subsidies available to homeowners or apartment owners who wish to install energy saving devices. Energy labels on major energy using appliances have only become mandatory in a few countries. The Common Market which, unlike the IEA, possesses some supranational legislative capability, has discussed standards on all heat producing devices for buildings and has begun a prototype support campaign to encourage innovations in buildings that save energy. A list of actions in the buildings sector covering IEA countries is shown in Table 9.

One effective technique employed in Sweden, Denmark, and the U.K. is the energy audit. An engineering or consulting firm prepares a study of an industry, such as bricks (the U.K.), or auto painting (Denmark), which tells policy makers what progress can be expected in that industry. More specific audits of individual firms are used by other firms to indi-

TABLE 9. IMPLEMENTATION OF ENERGY CONSERVATION MEASURES IN THE RESIDENTIAL AND COMMERCIAL SECTORS

	FINANCIAL/FISCAL INCENTIVES								BUILDING CODES								APPLIANCE EFFICIENCY						INFORMATION/ADVICE/ASSISTANCE							FUEL UTILIZATION	
	TAXES	Removal of Sales Tax	Discount on Taxable Income	SUBSIDY GRANT	LOAN	Others		Job Creation and Conservation	NEW HOMES Federal	Local	EXIST. HOMES Federal	Local	Federal/Public Buildings	MAX. TEMP Air	Water	Prohibition of Bulk Meters	ENERGY LABEL Mandatory	Voluntary	STANDARD Federal	Local	MAINTENANCE	Others	INFO. SERVICE/ CAMPAIGN	PUBLICATIONS	SCHOOLS	VISIT/CONSULT	HOME ENERGY AUDITS	TV/MASS MEDIA	Others		
Canada	—	X	—	X	X			X	X		X		X	—	—	X	X	—	P	—	P		X	X	X	X	X	X			
U.S.	—	—	P	X	P	X		X	X	P	—		X	—	—	P	X	—	P	—	—		X	X	—	P	P	X			
Japan	—	—	—	—	—			—	—		—		P	—	—	—	—	X	—	—	—		X	X	P	—	P	X			
N.Z.	X	—	X	—	X			—	X		—		X	—	—	—	—	—	—	—	—	X	X	—	—	—	X	—			
Austria	—	—	—	X	—			—	X		—		X	—	—	—	—	X	—	—	—		X	X	X	—	—	X			
Belgium	—	—	—	X	—			—	X		—		X	X	—	—	—	—	—	—	X		X	X	—	—	—	—			
Denmark	—	—	P	X	—			—	X		X		—	X	X	—	—	X	P	—	—		X	—	—	—	—	—	P		
Germany	—	—	X	X	X			X	X	X	X	X	X	X	X	P	—	X	X	X	X		X	X	X	P	—	P			
Greece	—	—	X	—	—			—	—		—		—	—	—	—	—	—	—	—	—		X	—	—	—	—	X			
Ireland	—	—	—	X	—			—	P		—		—	—	—	—	—	—	—	—	—		X	—	—	—	—	X			
Italy	—	—	—	—	—			—	X		X		X	X	—	X	—	P	—	—	—		X	X	X	X	X	X			
Luxembourg	—	—	—	—	—			—	—		—		—	—	—	—	—	—	—	—	P		X	X	—	—	—	—			
Netherlands	—	—	—	X	—			—	X	X	—		X	—	—	—	—	P	P	—	—		X	X	X	X	—	X			
Norway	—	—	—	—	X			—	X		X		P	—	—	—	—	—	—	—	P		X	—	—	—	—	—			
Spain	—	—	X	—	—			—	X		—				—	P	P	—	—	—	—		X	—	—	—	—	—			
Sweden	—	—	—	X	X			—	X		X		X	X	—	—	—	—	—	—	X		X	X	X	X	X	X			
Switzerland	—	—	—	—	—			X		X		X	X	X	—	—	X	—	X	—	—		X	X	—	—	—	—			
Turkey	—	—	—	—	—			—	P		P		—	—	—	—	—	—	—	—	—		X	X	—	—	—	X			
U.K.	—	—	—	X	X			X	X		—		—	—	—	—	—	—	P	—	—		X	—	—	X	—	—			

(1) Guidelines or voluntary standards.
(2) Did not state whether mandatory or voluntary.

-: no measure
P: planned measure
X: existing measure

SOURCE: International Energy Agency, June 1978

cate savings opportunities. The Danish auto painting study analyzes why some firms use as little as two-thirds the heat of other firms; the U.S. cement study does the same for cement. These studies are especially helpful to smaller firms whose own expertise and budgets are limited. In Sweden, the audit extends to the single firm, and the consultant gives the firm concrete suggestions on ways to improve energy productivity. Even estimates of costs and savings are given.

The lesson here is that in the industrial area, information is a vital ingredient in energy conservation. Larger, energy intensive firms, whose collective consumption accounts for as much as 75 percent of a nation's industrial energy use, are usually well informed about conservation possibilities. Smaller firms, however, have far less access to information and expertise. Hence the government, often through a branch organization, acts to provide that information. In the United States, some electric utilities have become consultants for small power and light customers, often under pressure from local utility commissions.

On the whole, however, energy conservation *programs* in Europe are still in their infancy, as in the case of the U.S. My own studies suggest that Sweden has progressed the farthest, followed by Denmark, Holland, the U.K., and the U.S. The IEA notes that most of the efforts are underfunded in relation to stated goals. Of course, it is hard to assign "saved" Btu's to specific programs when higher energy prices are themselves in large part responsible for changes in consumption patterns. On the other hand, aggressive energy standard programs, as have been implemented in California or Sweden in the residential or commercial sectors, do lead to identifiable savings in so far as structures (or appliances) are built to lower life cycle costs because of the standard. What all countries lack, however, is follow-up measurement—are the low energy houses in Sweden, Denmark, or California performing in fact as well as they were designed? Answering that question will not be inexpensive or easy but is very important for planning future conservation programs and even supply development. Unfortunately, few European countries have integrated detailed data on housing stock, industrial plant, or automobile stock into energy models or into the energy data books of national agencies. The result has been a great degree of difficulty in developing conservation programs or allocating resources, since so little appears to be "officially" known about consumption patterns. This view was constantly expressed in my discussions with officials in transnational organizations in Europe and in individual countries. In this respect the tremendous energy consumption data effort underway in the U.S. is admirable and will make an important contribution to our conservation planning while Europeans still argue over whether retrofit insulation is profitable in existing homes and apartments.

Life Style

While we have not treated life style explicitly, it is clear that this factor does enter into explaining differences in energy use patterns among countries. For the energy conservation planner wary of establishing normative conservation goals or standards, the issue of life style may be unwelcome. Nevertheless it is important to use our observations of other countries in an attempt to understand the possible couplings between energy conservation and life style.

Quantitatively there are two aspects of life style that bear directly on energy use: the mix of nonenergy goods and services, aside from energy itself, demanded by consumers, and the mix of key energy intensive activities that interact directly with energy. To the latter group belong indoor temperatures, patterns of auto ownership and use, land use patterns, appliance ownership, vacation and travel habits, and ownership of second homes or boats. The U.S., Canada, and Sweden tend to have the greatest energy demands arising from these patterns, while the remainder of Europe remains considerably "behind" the U.S. but is narrowing the difference somewhat.

It is hard to label activities such as living far from work as "wasteful"; yet, it is important to investigate why people live and work where they do; why they may evacuate cities on weekends for summer homes; why they prefer detached single-family dwellings to apartments. For example, most countries allow homeowners to deduct mortgage interest payments from taxes, an important subsidy for homeowning, especially in high tax countries like Sweden. Moreover commuters in Sweden can deduct the cost of the monthly bus pass from income, and those who can prove that driving saves one half hour (each way) compared to mass transit can also deduct the full cost of driving. These "life style" subsidies may be justified on social grounds, but they have a measurable impact on spreading people out, which, in turn, tends to increase energy use.

Should any country "embrace" another country's life style for the sake of saving energy? Probably not. However important the connection between life style and energy, there are so many conservation opportunities that involve technology or minimal behavioral adaptation to higher energy costs that we may not *need* to consciously live like other peoples just to save energy. However, understanding the energy implications of alternative patterns of consumption and location certainly would illuminate options for society. Thus the energy comparison of Mora, Sweden, and New Ulm, Minnesota, created great interest in trying to quantify the energy implications of perceived differences in life styles in the two countries. In this case the market basket differences probably have less to do with observed differences in energy use than the life style (or technical) differences in direct consumption habits.

While little data yet exist that allow general conclusions to be made about energy and life style, details from the Swedish-American comparison, this study, and other work support some important tentative findings.

1. The greatest differences in driving habits arise in the use of the auto for short trips, far more prominent in the U.S. Commuting via auto is gaining, however, in all countries, and load factors are low, partly because people living in clustered areas are still riding mass transit. Greater distances in the U.S. affect distance to work, but do not account for the significantly greater distances traveled. Indeed, distance per car per year (Table 5) varies far less across countries, suggesting that it is the ownership of a car that sets off life style changes leading to increased driving nationally.
2. Land use planning influences life styles and energy use considerably. As people spread out into suburbs, often aided by government home-building subsidies, cars become a vital link to shopping and services. Still, zoning in Sweden allows some services to be "built-in" to residential areas, while in the U.S. the great suburbs seem to isolate residences from services.
3. The low relative cost of scheduled and charter air flights in the U.S., compared to Europe, offers an energy intensive but time saving alternative to auto vacation travel. In Europe, low cost charters have gained immensely in popularity, at least in the Scandinavian countries; but in most places the auto seems to dominate vacation travel, causing immense traffic problems never seen in this country. Additional studies should be made to compare patterns and costs of auto use in Europe and the U.S. Rail travel is still important for vacationing and even much intercity business travel in Europe because of high density. Density clearly helps mass transit.
4. Whole house heating is only saturated in the U.S., Canada, and Scandinavia. One OECD study suggests that affluence in Central Europe will support significantly greater demand for heat. Similarly, appliance use will grow. Auto ownership is still far from saturation in Central Europe and is growing dramatically. Whether autos and appliances will ever approach those in America in size is unclear. If European countries begin now with appliance standards, new large devices not yet in place could be significantly more efficient at similar levels of use and saturation than their American counterparts of the 1950s and 1960s.
5. Other life style aspects of living patterns remain to be understood vis-à-vis energy use. For example, what is the overall impact of commuting to second homes in countries like Sweden, Denmark, or France? Does the fact that Americans move every six years (on the average) inhibit our ability to design communities and residences for long range resource costs? Will increases in affluence in Europe lead to the "Americanization" that is observable in Sweden and Denmark?

Quantitatively it is possible to separate effects of life style from energy comparisons by concentrating on the use of heating, autos, and appliances. Whether life styles directly affect the intensities of *devices*, which can be affected by policies and prices, is unknown. In any case, we know that life styles do affect energy use, and we know that these structural effects are apparent in a few important areas. These differences account for a significant amount of the differences in energy use between North America and Central Europe (Scandinavia being intermediate). Since conservation affects mainly intensities, we can safely say that a great deal of conservation can be decoupled from life style issues, while further reductions in overall energy use might come about through key life style changes in the U.S. Whether these changes themselves would occur is another matter worth discussion elsewhere.

Summary and Conclusions

What have we learned from surveying energy use in industrialized countries? First, there is no question that other nations use energy in a variety of ways that are often more efficient than our own techniques, though the reverse may be true as well. These technical differences account for a major part of the differences in energy use among countries. Energy prices (or taxes), higher in most European countries (and in Japan) than in the U.S. or Canada, have played an extremely important role in bringing on these energy saving technologies. This factor is too often ignored by those who cite other countries as examples of energy conserving societies relative to our own. But we can say without doubt:

> International comparisons of energy use show that there is much technical flexibility and conservation potential within present U.S. energy use patterns, and in other countries, provided that economic incentives and time are allowed to play a role.

Energy use policies per se are of a secondary nature in the establishment of today's practices, but policies will be more important in the future. Energy saving building codes are the most important of these but are themselves only truly significant in Scandinavia and perhaps now part of the U.S. Most European countries plan to introduce or stiffen building codes, but the Scandinavians are by far the leading practitioners of energy efficient buildings.

Life style, coupled to energy through the standard and size of homes and the nature and extent of personal transportation, plays an important role in the differences in energy use, to some degree among the higher energy users (Scandinavia, North America) and to a greater degree among other wealthy countries. While the coupling between life style and energy must be considered in any energy policy, it must be stressed that energy has rarely if ever been a consideration in the formulation of transportation or housing policies in so far as style and other nontechnical factors are concerned. The fact that social policies may even encourage energy consumption from a structural point of view while encouraging conservation from a technical point of view—a situation already brought about on the structural side in the U.S. twenty years ago—suggests that energy will remain the tail of the dog but never the whole animal. Certainly energy and nonenergy policies need close coordination, but Swedish experience suggests that such is not always simple. Ultimately, however, energy policy should be able to fit in with other important goals as we gain more knowledge of the link between energy and everything else.

It remains to pass judgment, however, on an important issue raised by international comparisons of anything. Sweden is not the U.S. (and vice versa). Are our comparisons meaningful in any sense? Does the small size of Sweden's homogeneous population, cultured in a tradition of political trust not common in the United States, allow institutions and people to function differently in reacting to a national problem? Does the lack of a class of poor people allow the Swedish government to carry out stern (but necessary) measures related to energy while we legitimately worry for years over the impact of any policy or economic move on our large class of poor? These are the kinds of issues that most often surface when any mention of energy policies or experiences in other countries occurs. It is often assumed that such issues make really detailed international comparisons worthless.

My own counter to the negative implications of such questions is that the goal of our work is to find ways in which we can economize on the use of energy and other resources. The evidence is clear that in certain important areas great economies have already been achieved, both in the U.S. as well as abroad. To argue that experience abroad cannot be studied and tailored to U.S. conditions is in a sense to assume that there is something in the U.S. system that makes economic use of resources difficult or impossible. Historically America has become more economic with resources over the long term. To insist that new economies that are technically possible, economic, and already in place in other parts of the world are somehow unachievable because "we are not they" seems to me incorrect. Rather it behooves us to observe what we—and they—are doing as a means of improving on what we would otherwise do in the future—and thereby conserve energy. After all, energy conservation, according to economist Kenneth Boulding, is just "thinking before using energy."

Robert F. Hemphill, Jr.

4

Energy Conservation in the Transportation Sector

Introduction

From the first months of the embargo in October and November of 1973, attention has been focused on the transportation sector as a major actor—to many, a real villain—in the energy conservation drama. There is plentiful justification for this attention. If we have an oil problem, as we clearly do now and are likely to in the future, transportation is key: while it uses only 26 percent of our total energy, all but a tiny fraction (less than 1 percent) of this is oil. The focus becomes even sharper when one examines the make-up of the transportation sector. The automobile has long been recognized as a principal element in our economic system; some estimates are that one of six jobs depends on it. Our social fabric—drive-ins, car washes, gas stations on every other corner, beltways, suburban living in general—and our popular culture —hot rods, recreational vehicles, *Smokey and the Bandit, The Kandy-Apple Tangerine Flake Streamline Baby*—also testify to the particular significance of this handy piece of machinery. Yet the passenger car is a voracious consumer of petroleum.

ROBERT F. HEMPHILL, JR. *is Deputy Assistant Secretary for Planning and Evaluation, U.S. Department of Energy. Mr. Hemphill has held energy-related positions with the Department of Health, Education, and Welfare and the Office of Management and Budget. Before returning to Washington in his present capacity, Mr. Hemphill was Deputy Director, Energy Conservation Policy Center, Carnegie-Mellon Institute of Research.*

Fig. 1. 1976 Distribution of Transportation Energy

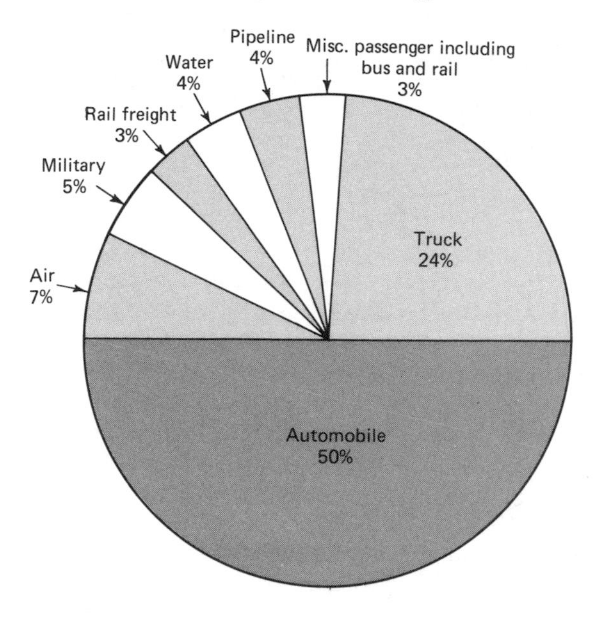

As such, it has been vilified as representing our most obviously wasteful and selfish tendencies, and countless hours of public policy and program attention have been devoted to the "gas guzzler" and what to do about it. The principal argument of this chapter is that there has been more than enough debate about the efficiency of the automobile as a vehicle; sufficient action is now underway to ensure the demise of this inefficient species and its replacement with a smaller, lighter, more efficient, and perhaps differently powered substitute. There are now enough laws on the books, or pending in Congress; new bureaucracies have been established to oversee these laws; and engineers are at work in the auto industry to comply with their provisions. While there are difficulties in forecasting what the final improvement will be—it may not be as great as the numbers appear—major improvements are coming. And the same essentially holds true for the *vehicle* efficiency of the other prime movers in the transportation sector. Interest should now be focused on the less analyzed area of how to use the vehicles efficiently—questions essentially of load factor and systems efficiency. This is an area of great payoff; little has been done to realize that payoff. If we are to make major new efficiency improvements in the transportation sector, they must next come from improving in-use efficiency. Vehicle miles per gallon has been or is being taken care of. Passenger miles per gallon, or ton-miles of freight per gallon, remains virtually unaddressed by public policy in any significant or effective way.

How Did We Get Where We Are?

Any analysis of historical trends in transportation energy use reveals conclusions that are in one sense surprising and in another perfectly reasonable. Basically, we have gradually shifted our transportation "business" to increasingly less energy efficient vehicles and modes, and we have increased our use of transportation services at almost the same rate that GNP has increased. Over the last fifteen years, freight movement grew at a rate 11 percent lower than real GNP growth, but passenger travel grew 26 percent faster. In many cases, we have traded more energy use for better service. But this has not necessarily been an immoral or inefficient (in the purely economic sense) decision, for two reasons.

First, energy costs are only a small percentage of transportation expenses. This is an even bigger barrier to conservation in industry, since energy costs are an even smaller part of the cost of doing business. According to the Federal Highway Administration, less than 20 percent of the cost of owning and operating an automobile is directly attributable to gasoline costs. Table 1 shows comparative figures for other modes. Note that energy costs are all less than that for automobiles.

Second, the real cost of fuel, as Pindyck points out effectively in Chapter 2, declined from 1950 to about 1970. In such a situation it makes clear economic sense to substitute less expensive inputs for more expensive inputs (e.g., labor) wherever possible. Moreover, in constant 1973 dollars, the price of a gallon of gasoline rose about ten cents in the year following the embargo, but actually dropped over the succeeding three years.

The trend toward inefficiency in automobiles has been particularly pronounced. Since these vehicles consume such a large share of the transportation sector's energy, this should have been, one asserts with the

TABLE 1. FUEL COSTS AS A PERCENTAGE OF OPERATING COSTS

Mode	*Percent*
Rail	7-8
Truck	15-19
Air	12
Pipeline	15
Intercity bus	2-3
Urban bus	1-2

acute vision of hindsight, a cause for some public concern. The fuel economy of new cars decreased steadily from the mid-1960s to 1974, hitting an all-time low of 13.9 miles per gallon. This decline is reflected in the average fleet fuel economy calculations, and both are depicted in Figure 2. Three factors are generally cited to account for these changes.

Increased Weight—Fuel consumption is almost directly proportional to weight—hardly a surprising conclusion to draw from the empirical data when one thinks in the basic terms of mechanics. Energy is used to effect a change in momentum (that is, mass times velocity), and the essential formulas used to make these calculations are all first-degree equations. As engines grew in size and options were added to automobiles over the years—automatic transmissions, power steering, power brakes, air conditioning—weight was added as well, and fuel economy declined.

Increased Comfort and Performance—Additional options not only add weight, they have two other impacts on fuel economy. First, in some cases they decrease the efficiency of the energy conversion system of the

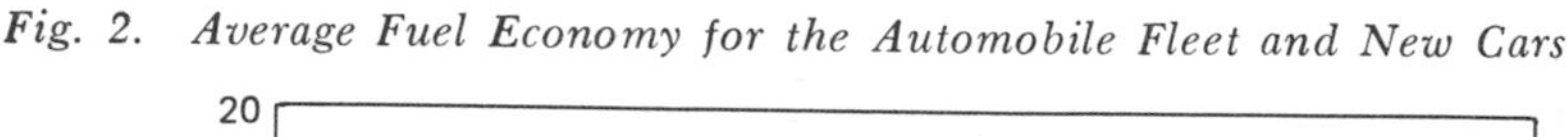

Fig. 2. Average Fuel Economy for the Automobile Fleet and New Cars

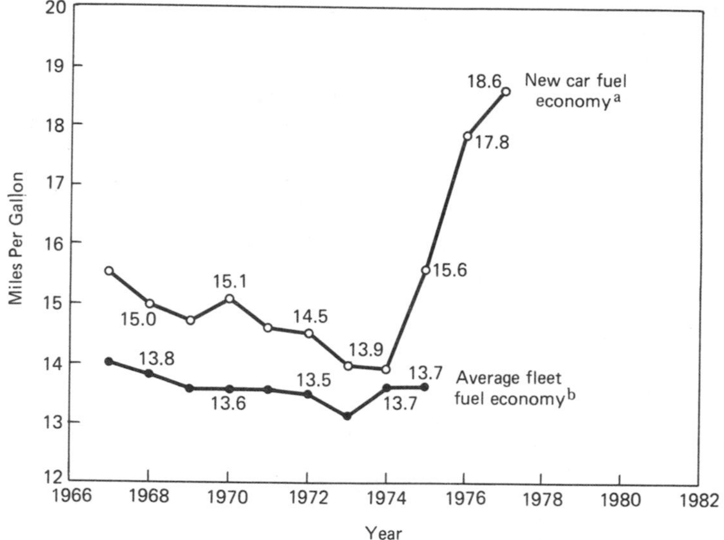

[a] Sales-weighted city/highway combined fuel economy by model year based on post-1974 test procedures.

[b] Average miles traveled per gallon of fuel consumed as reported annually by the Office of Highway Planning, FHWA.

car; the best example of this is the automatic transmission. Second, they drain power from the engine to perform these service functions, e.g., air conditioning. In the latter case the combined weight addition and power cost of air conditioning has been estimated by EPA to work out to a 9 percent average fuel economy penalty. This has led to so far unsuccessful attempts by several congressmen to ban air conditioning in new cars. Despite these penalties, the percentage of new cars sold with air conditioning is above 80 percent.

Tightened Auto Emission Standards—An extraordinary and contentious debate has raged among the auto industry, the Congress, the environmental community, pollution control equipment manufacturers, and several interested elements of the federal government on the questions of the actual impacts of fuel economy to date of the Clean Air Act auto emission standards, whether these impacts were avoidable had the industry used better technology, and the likely future impacts of the increasingly stringent standards found in the law. To attempt even to lay out fairly the arguments and associated analysis on this issue would require far more than a chapter. Let it only be stated as the author's best judgment that:

—for better or worse, avoidable or no, the techniques chosen by the auto makers up to 1974 to meet increasingly tough standards—principally spark retardation and exhaust gas recirculation—have had negative impacts on fuel economy. In 1975, as major manufacturers adopted catalyst control technologies, fuel economy rebounded upward.

—there is some level of environmental stringency that can be established in the future that will eliminate diesel engines as power sources in automobiles. The levels, the fuel economy impacts, and the national costs associated with such a large potential change are extraordinarily difficult to predict. Most such predictions are little more than speculation.

Two other factors have contributed to the growth in motor gasoline consumption. First, there has been an enormous growth in light truck sales; these now account for over 25 percent of new vehicle sales. Trucks do have capabilities to carry weight, but they are most often used as passenger car substitutes—with a 15 percent lower average fuel economy.

Second, the use of public transit has steadily declined, in terms of passenger miles, at slightly more than 2 percent a year, from a high of 3,254 million passenger miles in 1945 to 2,021 million in 1977. It is now only just holding its own on an aggregate basis. As Table 2 indicates, the number of public transit vehicles has fallen at the same rate, at least until the last three years, with the trolley/light rail class taking

TABLE 2. TRANSIT PASSENGER VEHICLES OWNED AND LEASED

	Railway Cars					
Calendar Year	*Light Rail*	*Heavy Rail*	*Total Rail*	*Trolley Coaches*	*Motor Buses*	*Total Revenue Vehicles*
1945	26,160	10,217	36,377	3,711	49,670	89,758
1950	13,228	9,758	22,986	6,504	56,820	86,310
1955	5.300	9,232	14,532	6,157	52,400	73,089
1960	2,856	9,010	11,866	3,826	49,600	65,292
1965	1,549	9,115	10,664	1,453	49,600	61,717
1970	1,262	9.338	10,600	1,050	49,700	61,350
1971	1,225	9,325	10.550	1,037	49,150	60,737
1972	1,176	9,423	10,599	1.030	49,075	60,704
1973	1,123	9,387	10,510	794	48,286	59,590
1974	1,068	9,403	10,471	718	48,700	59,889
1975	1,061	9,608	10,712	703	50,811	62,226
1976	963	9,714	10,720	685	52,382	63,787
1977 *	992	9,639	10,674	645	51,968	63,287

* Preliminary

the heaviest casualties. The less efficient automobile has triumphed as public transit vehicle use declined.

What's to Be Done about the Gas Guzzler?

The fascination with the automobile as culprit in the energy policy debate has had a certain anthropomorphic tenor to it. If the private car has become big, heavy, and inefficient, either through some dimly understood biological process or the venality and lack of public spirit of Detroit, then the solution is clearly to make the vehicle more efficient. That is certainly in part correct, and policy-makers (including the author) have taken part in a campaign which is likely to be successful in accomplishing that aim, if other public goals do not interfere. A brief inventory of the existing and seriously proposed laws and programs is useful at this point.

MANDATORY FUEL ECONOMY STANDARDS

Passed as part of the Energy Policy and Conservation Act of 1975 (EPCA) (P.L. 94–163), the law, as currently implemented by the secretary of transportation, sets the following requirements for the average of each manufacturer's new car fleet:

Model Year	*MPG*
1978	18
1979	19
1980	20
1981	22
1982	24
1983	26
1984	27
1985 and beyond	27.5

Serious financial penalties are to be levied on any manufacturer who fails to meet these standards, and the major auto makers have indicated their intent to hit these targets.

The secretary also has the administrative authority to modify the 1985 and later year standards, but such a decision is subject to veto by either house of Congress if he increases it above 27.5 or decreases it below 26.0. In the rule-making which established the 1981-84 standards, Secretary Brock Adams in fact gave notice of his eventual intentions to tighten these standards. Future levels are difficult to predict but figures as high as thirty-five to forty miles per gallon have been discussed.

FUEL ECONOMY TAXES AND REBATES

President Carter, in the April 20, 1977, National Energy Plan, proposed to tax inefficient autos and to provide rebates to the purchasers of efficient autos as a further stimulus to the production of efficient cars. This program passed the House in modified form, although the rebates were deleted because of concern about foreign trade impacts, and the tax penalty schedule was made somewhat less ambitious. The Senate passed a measure to ban low-mileage autos rather than tax them, and the issue now has been dropped.

AUTOMOBILE ENGINE RESEARCH AND DEVELOPMENT

For several years the federal government has been sponsoring research on alternative automobile engines that would have as good or perhaps better fuel economy and as low emissions as the advanced gasoline and diesel engines likely to be in production in the late 1980 to early 1990 period. In addition, the two candidate engines—the gas turbine and the Stirling—should, if the research program is successful, both be able to meet tougher environmental standards than currently exist and to use a broad range of liquid fuels. The budget for this program is likely to increase to more than $50 million per year by 1980 as prototypical cars are built and tested. The Department of Energy also manages a program of research, loan guarantees for manufacturers, and demonstration purchases designed to develop a competitive electric or hybrid vehicle. The cost of this program is likely to be in the $30 to $40 million range by 1980, with fleet deployment not expected until the early 1990s.

INFORMATION

The EPCA, in addition to setting fuel economy standards, requires that fuel economy labels be displayed in every new car, giving both the fuel economy of that particular model and the range of fuel economies found in cars of the same size class. It also mandates the publication and distribution of a booklet which includes all the fuel economy ratings for cars sold in the U.S. Some 20 million copies of this *Gas Mileage Guide* are distributed annually to all auto dealers in the country, who in turn are required to make a copy available to each prospective customer.

In sum, almost all of the generic federal policy tools—regulation, tax incentives and penalties, R&D, loan guarantees, information—have been either proposed or implemented with the direct objective of improving the efficiency of the vehicle itself. Fuel economy is already up dramatically and should go higher. Weight reductions are already being made; friction losses are being attacked; gasoline engines are improving; diesel engines are entering the fleet. Table 3 lays out the Department of Trans-

TABLE 3. FUEL ECONOMY IMPROVEMENTS AVAILABLE IN THE 1985 NEW AUTOMOBILE FLEET

Modification	Δ *MPG* %
Weight Reduction (1980–1985)	11
Engine size reduction	3
Dieselization (or equivalent) *	6
Improved transmissions	8
Lubricants	2
Rolling resistance	3
Accessory Load Reduction	2
Reduced aerodynamics	4

* 25% penetration of 25% more efficient engines.

portation's estimates for the sources and magnitudes of these improvements. And one is strongly tempted to conclude that enough is enough: put the programs in place and give the industry and the consumer time to react.

How Can We Make Auto Use More Efficient?

In comparison to the stress on vehicle fuel economy, efforts to change the way we use automobiles have been limited, poorly received, and generally ineffective.

Autos are used rather prodigally in this country:

1. Fifty-eight percent of all auto travel is done on urban streets; more than half of all trips taken are of five or fewer miles; the average household makes 3.8 auto trips per day.
2. Trips involving 100 miles or more have load factors of two persons per trip; urban trip load factors are substantially lower, on the order of 1.4 persons per trip.
3. One-third of all automobile use is in getting to and from work, an average daily round-trip of 18.8 miles. And this travel is largely done alone: four out of five auto commuters drive alone to work in their five or six passenger autos.
4. Research indicates that through simple, noncostly changes in driving technique, the average motorist could improve his fuel economy by 10 to 20 percent. Such improvements, however, are singularly difficult to target by specific public policies.

Many thoughtful studies have concluded that the single-occupant commuter offers the best target for energy conservation because of his routine, fixed destination, route, and schedule. The remaining travel categories—business-related travel (8 percent), family business (25 percent), and social and recreational travel (34 percent)—do not lend themselves nearly so easily to increasing the occupancy rate as does commuting. Studies indicate that only 20 to 25 percent of commuters are really prohibited from carpooling because of odd schedule, job location, or home site situations; yet this potential is hardly tapped. If only one of every two single-occupant vehicles had two persons in it coming to work rather than one, savings of 200,000 barrels per day of oil would result. And no new vehicles or capital facilities are needed to bring about this change.

Are carpools a "better" solution than mass transit, from an energy standpoint? The conclusions of the best recent analysis of the Btu per passenger mile ratings of each of the principal urban passenger carriers, including energy for propulsion, vehicle and guideway construction, and roundabout trips, are presented in Table 4.

TABLE 4. BTU'S PER PASSENGER MILE FOR VARIOUS URBAN TRANSPORTATION MODES

Mode	*Btu per Passenger Mile*
Dial-a-ride	17,230
Single-occupant auto	14,220
Average auto	10,160
Heavy rail transit (new)	6,580
Carpool	5,450
Light rail transit	5,060
Commuter rail	5,020
Heavy rail transit (old)	3,990
Bus	3,070
Vanpool	2,420

Examination of this table brings home two factors:

1. Buses and vanpools are big winners if the game is scored purely on energy intensiveness and conservation potential, and
2. Carpools are better than new heavy rail and about equal to light rail and commuter rail.

Analysis of a single variable does not always make for high-quality policy recommendations. Urban transportation must also be affordable, and new heavy rail transit systems appear increasingly not to meet this criteria, at least in the minds of their riders. Put another way, there is little sense in saving energy at any cost, and some new transit proposals, especially new heavy rail systems, are quite expensive. An analysis done several years ago of the Washington, D.C., Metrorail system concluded that the system would assuredly save energy, diverting about 5 percent of automobile commuter traffic, but at a present-value cost of $313 per barrel of oil saved. There are good reasons for transit systems, and increased use of existing systems should surely be a part of any coherent energy conservation strategy; but new, expensive rail systems appear to make little sense, in pure energy conservation terms, in any new city.

The same is not true for the bus, carpool, and vanpool modes. These share the advantage of being relatively less capital-intensive, of not being tied to fixed guideways, and of being able to expand or contract to meet new demand fairly easily. The major disadvantage of bus systems in today's cities is, of course, the dispersed, low-density residential patterns that have grown up. This means a longer amount of time must be spent picking up riders, compared to line-haul times, and tends to discourage ridership. Moreover, the fact that a paid driver is used, that service must be maintained during nonrush periods, and that other costs (e.g., parking during nonduty periods) are borne directly by buses but only indirectly by cars and vans, combine to drive transit prices above levels where riders with other options will voluntarily bear the full cost of the services, especially given the relative inconvenience of taking the bus. The result has been the widespread local subsidization of transit lines, and federal capital (and recently operating) subsidies as well. Less than 10 percent of transit ridership in the U.S. currently receive no benefits from government subsidy.

There are two final disturbing notes to add to this brief discussion of mass transit, both having to do with ridership.

First, it is devilishly difficult to get people, especially the right people (solo-occupant auto commuters), onto transit systems. An elaborate transit system saves no energy at all without riders. A number of studies and experiments have found that attracting riders to bus systems of reasonable cost and standard service is very difficult; the auto alternative is simply too appealing, measured in terms of convenience, privacy, perceived cost (gas and oil expenses are only 17 to 20 percent of the actual costs of owning and operating a car), and personal control one can exercise. Buses just are not an attractive alternative to most commuters. And for the new rapid rail transit systems, there is some disturbing evidence that large numbers of riders are coming from buses and carpools, rather than from single-passenger autos, a perverse, energy intensive effect indeed.

Second, quick, efficient, attractive rapid rail systems may contribute to further dispersed living patterns rather than to more concentrated, and thus generally more efficient, urban areas. If one assumes that for the majority of middle-class commuters the most important commuting cost is time rather than money, then a rapid rail station on the Washington beltway (with a large parking lot, to be sure) allows one to live in a bedroom community twenty miles outside the beltway and commute daily to downtown Washington in little more *time* than a nonsubway trip from upper Chevy Chase, ten miles from center city, would take. Again, in conventional energy terms this is a perverse effect, although it may be useful if we move to an energy system based entirely on decentralized solar energy with the associated need for acres of collectors.

The major government programs and proposals to encourage increased efficiency of auto use have either been only tangentially effective, or they have been rejected by the Congress. A quick summary of the major ones follows.

1. Raising the price of crude oil to world market levels, either through a decontrol/windfall profits tax arrangement or by a tax and rebate scheme,

has been proposed by various administrations in recognition of the need to stimulate more efficient use of petroleum and clean up the oil price regulatory system. Neither has been accepted by the Congress.

2. Contingent gasoline taxes were passed once by the House Ways and Means Committee in 1975 and roundly defeated on the floor of the House. They were again proposed by President Carter in the National Energy Plan and this time failed, by a two to one margin, even to make it through the Ways and Means Committee. Senate interest and action have been nil. And yet other industrialized countries sell gasoline at prices three to five times ours, with few noticeable ill effects.
3. Other forms of government action have been proposed and in some cases adopted, including various forms of demonstrations, economic subsidy, or preferential treatment for transit and carpool/vanpool arrangements, but their net impacts have been small in energy terms. Private auto disincentives or restrictions have been discussed, used in other places (e.g., Singapore), and are part of EPA's Transportation Control Plans, but have not been put into effect yet anywhere in the U.S. In fact, Congress repealed EPA's authority to require taxes on parking.

The issue of efficient usage remains a difficult one, but one which should be accorded top priority. The basic problem is that the painless solutions (carpools, vanpools, and bus transit) do not seem to be effective, and the painful ones (e.g., price increases or restrictions on the movement and parking of single-passenger autos) have no constituencies. Yet this is the transportation area where the largest savings are available at the lowest total social and economic cost.

Airplanes Are Here to Stay

In the intercity passenger travel market, the energy efficiency picture is somewhat gloomy. Only here, surprisingly, the villain is the airplane and not only the private car. Much work has been done on the relative energy use per passenger mile of the major competitors in the intercity market: autos, rail, bus, and air. While all average calculations are the captives of their assumptions as to trip length, load factor, vehicle configuration, and so forth, the estimates fall in the following general ranges:

Mode	*Btu's per Passenger Mile*
Bus	1,100
Auto	2,100-3,500
Rail	2,000-4,000
Air	5,000-10,000

The trends of utilization point in quite the opposite direction than the energy efficiency numbers would lead one to suggest they should go. As can be calculated from Table 5, air travel grew annually by 10 percent; auto travel grew 3.4 percent; rail travel declined 5.2 percent per year; and intercity bus travel grew by 1.9 percent. As a result, between 1960 and 1975 all the other modes had lost market share to the *least* energy efficient—air travel. Given the speed and convenience of airlines, there seems little chance of reversing or even damping these trends through any politically saleable public policy.

There is some potential, and some progress has been made in improving air's abysmal energy efficiency, both through vehicle changes and upgrading efficiency. The introduction of wide body aircraft, for example, has reduced Btu per seat mile ratios noticeably. NASA has embarked on a ten-year, $25 million per year program to create an even more energy efficient transport aircraft. The one notable exception to all

TABLE 5. INTERCITY PASSENGER MILES BY MODE OF TRAVEL, 1960 THROUGH 1975

	Automobile [a]	*Bus* [a]	*Total Motor Vehicle* [a]	*Rail, Revenue Passengers*	*Inland Waterway*	*Air, Public and Private*	*Total*
			Passenger Miles by Mode (in Billions)				
1975	1,164.0	25.5	1,189.5	9.7	4.0	148.0	1,351.2
1974	1,143.4	27.6	1,171.0	10.4	4.1	146.2	1,331.7
1973	1,174.0	26.4	1,200.4	9.3	4.0	143.1	1,356.8
1972	1,129.0	25.6	1,154.6	8.7	4.0	133.0	1,300.3
1971	1,071.0	25.5	1,096.5	8.9	4.1	119.9	1,229.4
1970	1,026.0	25.3	1,051.3	10.9	4.0	118.6	1,184.8
1965	817.7	23.8	841.5	17.6	3.1	58.1	920.3
1960	706.1	19.3	725.4	21.6	2.7	34.0	783.7
			Passenger Miles by Mode (Percent)				
1975	86.14	1.89	88.03	0.72	0.30	10.95	100
1974	85.86	2.07	87.93	0.78	0.31	10.98	100
1973	86.53	1.95	88.47	0.69	0.30	10.55	100
1972	86.82	1.97	88.79	0.67	0.31	10.23	100
1971	87.12	2.07	89.19	0.73	0.33	9.75	100
1970	86.60	2.14	88.74	0.92	0.33	10.01	100
1965	88.85	2.59	91.44	1.91	0.34	6.31	100
1960	90.10	2.46	92.56	2.76	0.34	4.34	100

[a] Includes intracity portions of intercity trips. Omits rural to rural trips, strictly intracity trips with both origin and destination confined to same city, local bus or transit movement, nonrevenue school and government bus operations.

this, of course, is the SST which is roughly one-fourth as efficient as a 747. But that, at least, is a French and British effort, not one endorsed by the United States, although partially condoned by our grant of Concorde landing rights.

The move to increasingly free airlines from government regulation by introducing competitive forces and thus economic efficiency into the area should also have a salutary effect on energy efficiency. Analyses indicate that unprofitable schedules and service will be curtailed and profitable service expanded; the long-run effect will be higher load factors and thus less energy used per passenger transported a given distance. Whether this will even come close to equaling the increased energy likely to be consumed in transporting increased aggregate numbers of passengers is very doubtful, but it is probably about the best we can hope for.

The Freight Sector Is in Fairly Good Shape

The data on the energy impacts of moving the nation's goods indicate a set of conditions somewhat different and more encouraging than those for passenger transportation. The energy efficiencies, with the major exception of air, fall within a fairly narrow range. The chart below lays out the generally acknowledged ranges of values for these efficiencies, with specific figures highly dependent on vehicle configuration, routing and trip length, load factor, commodity transported, and so on:

Mode	*Btu per Ton-Mile*
Water	100-500
Pipeline	100-1,300
Rail	200-1,000
Truck	1,100-2,000
Air	8,000-25,000

The historical data on relative market shares indicate some interesting shifts over time, but they are both mixed with regard to energy efficiency and not of the order magnitude of those in the intercity passenger market —see Table 6. In summary it should be noted:

1. The more energy efficient modes, water and pipeline, moved more than half of the freight in 1960; they have picked up almost two points of market share since, with pipelines gaining at the expense of water. It is perhaps ironic that the increase in one of the most energy efficient modes (pipelines) is at least in part due to increasing energy shipments and consumption.
2. Air cargo increased at an annual rate of 12.8 percent, more than twice as

fast as the next highest growth area (oil pipelines at 5.6 precent), but still less than 1 percent of the nation's freight moves by air. Given the energy intensity of this mode, this is just as well.

3. Rail and truck freight both increased in the aggregate, but truck freight grew more rapidly (2.9 percent) than rail (1.8 percent) and thus increased its market share by 1 percent.

Are there any real needs or opportunities for major conservation efforts in this sector? To answer this, one needs to examine, once again, both the efficiency of the vehicle or prime mover and the system efficiency.

One-fourth of transportation energy goes into trucks, compared to one-half for cars, even though there are four cars for every truck in the U.S. Trucks are, obviously then, more intensively used and have a lower miles per gallon rating than cars. The average tractor-trailer combination traveled 50,000 miles in 1975, compared with 10,000 miles for a passenger car. Moreover, the population of light-duty trucks (less than 10,000 pounds) has been increasing; a 1976 survey of twenty-one standard metropolitan statistical areas (SMSA) showed that trucks made up 8 percent of the daily urban commuting fleet.

Government action to date dealing with the efficiency of the vehicle has been limited to light-duty trucks. Vehicles in this class have been included under the EPCA-mandated fuel economy standards, although the target they must meet is lower. The 1981 standard is eighteen miles per gallon, compared to twenty-two for autos in the same year. There has been some discussion—mostly apprehensive discussion in the industry—about federal mileage standards for heavier vehicles. However, the variety of duty cycles and specialized purposes make this an option of extremely dubious administrability, and it has never been seriously considered politically. In fact, the truck manufacturers have been very responsive to the need for better fuel economy and have introduced various options (derated engines, air flow deflectors, increased use of diesel engines, clutch-actuated fans) to improve this. Truck operators pay a fair proportion of their costs for fuel, at least compared to the average manufacturing firm, and have in general responded well to these more efficient vehicles. In short, the combination of market forces for the heavy-duty trucks and standards for the light-duty trucks is about as much as can reasonably be done for the efficiency of the freight-moving highway vehicle. The efficiency of the prime mover for rail and water is already so high that technical improvements, while worthwhile, will have little overall energy impact.

Systems efficiency is a more difficult question. In one sense, the freight moving system is already reasonably efficient, with 80 percent of the tonnage moving by the three most efficient modes (see Table 6). The counter argument is that the complex economic regulation of the freight

TABLE 6. FREIGHT TON-MILES, 1960 THROUGH 1975 (MILLIONS)

	1960	*1965*	*1970*	*1971*	*1972*	*1973*	*1974*	*1975*
Trucks	285,000	359,000	412,000	445,000	470,000	505,000	495,000	441,000
Class I rail	572,309	697,878	764,809	739,743	776,746	851,809	850,961	752,816
Air carrier								
Total	869	1,968	3,295	3,457	3,662	5,051	5,531	5,323
Certificated	749	1,670	3,010	3,151	3,403	4,759	5,251	5,061
Supplemental	120	298	285	306	259	292	280	262
Water transport								
Total	NA	752,044	914,755	909,194	942,235	942,913	941,227	NA
Inland waterways including Great Lakes	220,253	262,241	318,560	316,030	338,693	358,222	354,882	343,000
Total domestic system	NA	489,803	596,195	593,164	603,542	584,691	586,345	NA
Oil pipeline	228,626	306,393	431,000	444,000	475,800	507,000	506,000	510,000

system maintained by the Interstate Commerce Commission has troublesome energy consequences in two ways. First, the regulatory system has placed barriers in the path of the railroads in trying to compete with motor carriers. The eleven-year decision process on the Rock Island Line merger application is frequently cited as a classic example. Second, within the sphere of motor carrier regulation, various rules (such as route or backhaul restrictions) often make the truck owner operate in an energy inefficient manner.

Some limited deregulation has been proposed in this area, principally to stimulate economic efficiency rather than for energy purposes. The premise behind such proposals is simple; if trucking operations are made more competitive and thus more economic, dollars will be saved, and some of those will be fuel dollars. Unfortunately the analytical tools to deal with the energy use of the freight system are not as highly developed as one would like. In fact, an early analysis of one deregulation proposal concluded that energy use would increase rather than decrease. The jury is essentially out on this, but the point to emphasize is that the total system is unlikely to change rapidly no matter what government policies are adopted. To a large extent, we are captives of our own infrastructure here, with dispersed pickup and delivery points, an existing stock of vehicles, and the associated business institutions and arrangements. And such policies as are proposed will no doubt derive basically from concerns about competition and economic efficiency, not energy. And that is probably as it should be.

Policy Options

Various estimates have been made of the potential savings that could be realized in the different portions of the transportation sector. In many cases these estimates are neither sophisticated nor reliable, for several important reasons. First, our data base, although probably better here than in other sectors, is not as good as one would like; moreover, our range of past experience, *even where we have good data,* is too limited to allow for valid predictions. For example, if we know in some detail that over the past five years, increasing the efficiency of a five-passenger auto by 10 percent has added 2 percent to its purchase price, we cannot logically assume this relationship will hold for a 50 percent increase in efficiency. Similar problems plague estimates on the conservation impact of price increases. Finally, in many cases, even the direction of the energy consumption change that will result from a policy action may not be as certain as one would hope. Several examples of this counter-intuitive effect have already been cited, and they derive from the simple fact that complex systems, such as those involved in energy use, have complex reactions, even to simple changes or stimuli.

TABLE 7. ENERGY SAVINGS SUMMARY (MILLION BARRELS PER DAY)

Sector	*1985 Baseline*	*Expected Saving*	*Maximum Additional Potential Saving*
Autos	5.2	.66	.9
Light trucks	2.0	.29	.13
Commercial trucks	1.6	.24	.13
Air carriers	0.9	.14	.06
Freight trains	0.4	.03	—

Given all the caveats, estimates still are needed to guide and order public policy. An optimistic set of such numbers has recently been prepared by the Office of Systems Engineering in the Department of Transportation. Table 7 indicates both the expected savings likely to result from policies or conditions already in place and the maximum feasible savings that are technically possible—although the public policies that would be required to bring about these savings are not specified.

This 900,000 barrel per day savings potential in the auto sector should be contrasted with estimates made by others of 67,000 barrels per day to 200,000 barrels per day, resulting from a specific policy, in this case a fifty cents per gallon gasoline tax. These differences in magnitude highlight the policy-maker's problem: even if we know what we want, how do we get there.

There are three conclusions that can reasonably be drawn from the material in this chapter.

1. There has been, if anything, *too much* attention focused on the efficiency of the automobile as vehicle and *not enough* on its efficiency as system. We now have a bountiful sufficiency of policy on auto fuel economy, but we need creative and/or courageous steps on auto use.
2. There is a real means and ends problem with auto use. There appear to be no policies for which the constituency and political will exists that would be effective in forcing increased auto use efficiency. Gasoline price increases would have some impact, but actual auto use restrictions of some sort are probably the only truly effective solution. This does not appear to be an idea whose time has come.
3. The potential for conservation in the other transportation sectors is not as substantial as that in autos, despite equivalent fuel use; it is not as easily obtainable; and much of the potential that exists will be realized by policies already in effect.

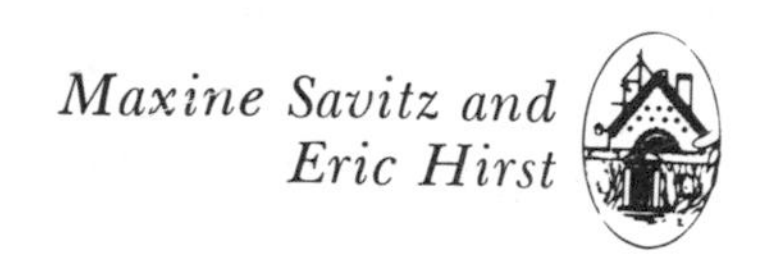

Maxine Savitz and
Eric Hirst

5

Technological Options for Improving Energy Efficiency in Residential and Commercial Buildings

Introduction

Built in an era of inexpensive and abundant energy, many of our nation's residential and commercial buildings are manifestly overlit, overheated, overcooled, and underinsulated. In 1976, energy demand for the nation's 74 million residential units and 1.5 million commercial buildings amounted to 26 quadrillion Btu's (quads). The nation's stock of residential and commercial buildings thus accounts for more than one-third of the total energy consumed in the United States. Because few, if any, energy conserving considerations were incorporated into their design, an estimated 40 percent—10.5 quads—of the energy used in these buildings is being wasted.

MAXINE SAVITZ *is Director of Buildings and Community Systems at the U.S. Department of Energy and editor of* Energy and Buildings. *Before joining the Department of Energy, Dr. Savitz was with the Energy Research and Development Administration.*

ERIC HIRST *is a research engineer in the Energy Division of Oak Ridge National Laboratory. Formerly he was Director, Office of Transportation Research in the Federal Energy Administration. Dr. Hirst has published many papers on energy issues.*

The residential and commercial sectors are thus prime targets for improving energy efficiency. In fact, the nation's ability to make the transition from profligate to efficient energy use will depend largely on whether we are willing to reorient some of the fundamental ways in which we think about and use energy in our homes, our stores, and our offices.

Thinking clearly about the present and planning intelligently for the future logically begins with a backward glance at energy use trends in the residential and commercial sectors. Against this historical backdrop, we can then explore a variety of options for improving energy efficiency that range from using commercially available devices as aids in changing our energy consumption habits to improving existing technologies and developing new technologies to experimenting with innovative approaches to energy use on the community scale. Before many of these options can be implemented, however, a variety of behavioral, institutional, and legal/regulatory barriers must be overcome.

After identifying the major barriers that impede the implementation of energy efficiency options in the residential and commercial sectors, we consider possible federal roles both to encourage and to ensure improved efficiency. We then investigate the energy and dollar saving implications of federal regulatory/incentive programs and research, development, and demonstration (RD&D) programs under several fuel-price scenarios at both the national and consumer levels. Our overall conclusions regarding the potential for and approaches to improving energy efficiency in residential and commercial buildings are presented in the final section of this chapter.

Trends in Energy Use

Before we can embark upon sound planning for the future, we need to understand and to place in perspective how we have used—and misused—our energy resources in the past. The following discussion of trends is based on *Recent Changes in Energy Use in Residential and Commercial Buildings* by Eric Hirst and Jerry Jackson.

Over the years, energy use has followed basically the same patterns in the residential and commercial sectors (see Figures 1 and 2). Both sectors' use of electricity, gas, oil, and other fuels (e.g., coal, liquefied gases) underwent two distinct stages from 1960 through 1975: a phase of steady and rapid growth (1960–1972)—averaging 5 percent in the residential sector and 6 percent in the commercial sector—followed by a phase of decline and instability (1972–75). Between 1975 and 1976, energy use then reverted to the earlier pattern, jumping 5 percent.

On a more specific level, energy use per household also peaked in 1972, declined through 1975, and then increased again in 1976—remaining below the 1972 peak, however (see Figure 3). The same pattern is

Fig. 1. Trends in Residential Energy Use

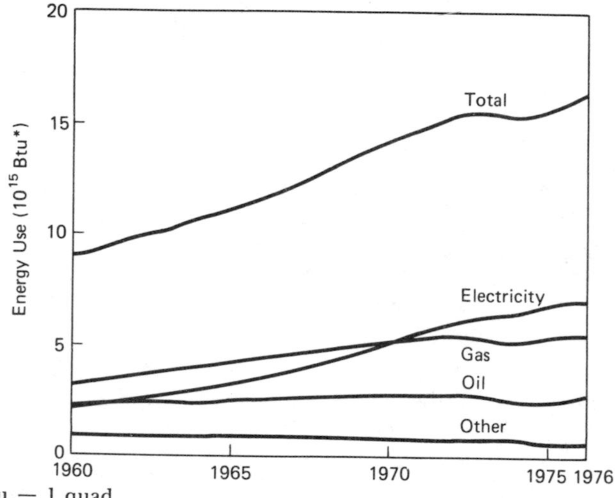

* 10^{15} Btu = 1 quad

Source: Eric Hirst and Jerry Jackson, *Recent Changes in Energy Use in Residential and Commercial Buildings,* Oak Ridge National Laboratory, February 1978.

Fig. 2. Trends in Commercial Energy Use

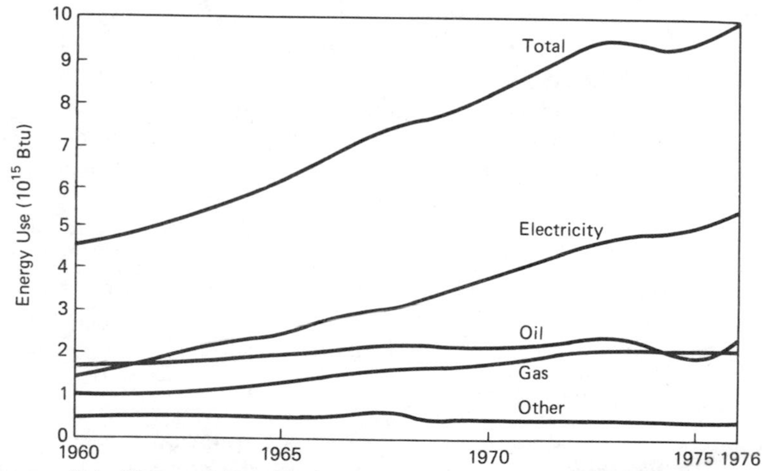

Source: Eric Hirst and Jerry Jackson, *Recent Changes in Energy Use in Residential and Commercial Buildings,* Oak Ridge National Laboratory, February 1978.

Fig. 3. Energy Use per Household

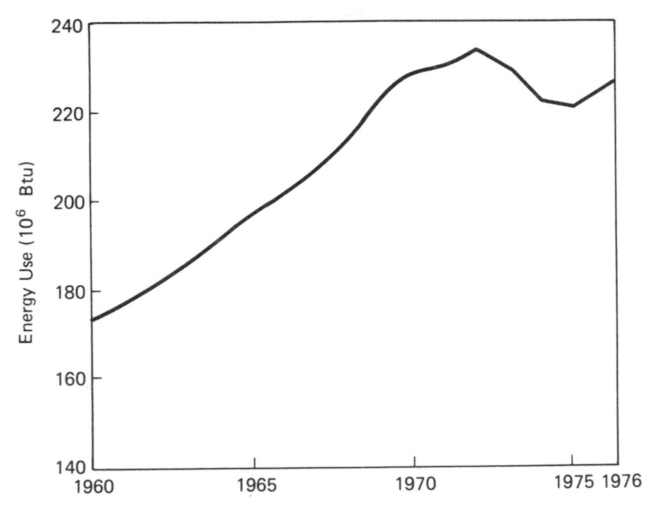

Source: Eric Hirst and Jerry Jackson, *Recent Changes in Energy Use in Residential and Commercial Buildings,* Oak Ridge National Laboratory, February 1978.

evident for energy use per square foot of commercial floor space and per dollar of economic activity (see Figure 4). Again, the peak occurred in 1972, with a subsequent decline through 1975 and a slight increase in 1976.

Trends in real fuel prices from 1960 through 1976 (see Figure 5) provide some insight into these trends in energy use. During the 1960s, electricity prices declined steadily and markedly, while gas and oil prices remained roughly constant. During the 1970s, however, all fuel prices rose dramatically: between 1972 and 1976, oil prices jumped almost 60 percent, gas prices increased more than 20 percent, and electricity prices rose 10 percent. This sudden reversal in fuel price trends undoubtedly accounted for much of the recent change in fuel use trends.

During the 1970s, economic growth also diverged sharply from its long-term historical trend. Between 1960 and 1973, real per capita income grew at an average annual rate of 3 percent. Between 1973 and 1974, per capita income dropped 2 percent; it then remained constant in 1975 and increased in 1976 to its 1973 level. These changes in economic activity are also likely to have influenced energy use patterns.

Options for Improving Energy Efficiency

Against the backdrop of these trends, the question now before us is: what can we do to conserve energy in residential and commercial buildings?

Fig. 4. Energy Use by Commercial Floor Space and Economic Activity

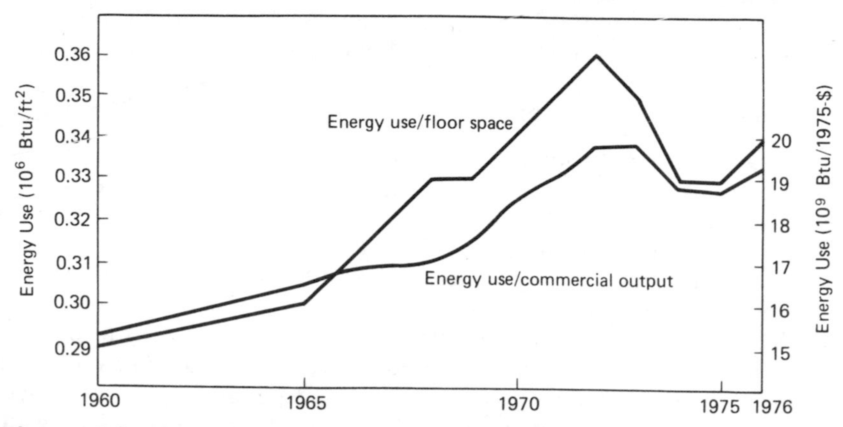

Source: Eric Hirst and Jerry Jackson, *Recent Changes in Energy Use in Residential and Commercial Buildings*, Oak Ridge National Laboratory, February 1978.

Fig. 5. Trends in Residential Fuel Prices

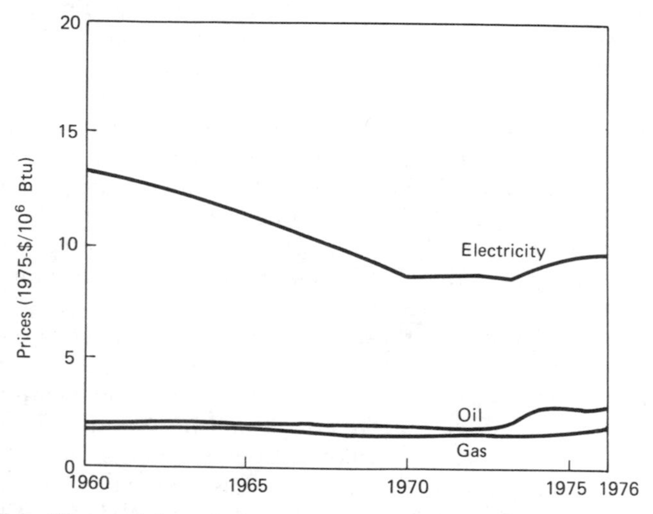

Source: Eric Hirst and Jerry Jackson, *Recent Changes in Energy Use in Residential and Commercial Buildings*, Oak Ridge National Laboratory, February 1978.

The variety of options that can be pursued to create an energy efficient built environment fall into three broad categories. First, we can realize immediate savings by using commercially available devices as aids in changing our energy consumption habits. Second, the improvement of existing technologies and the development of new technologies will result in substantial, long-term energy savings. And third, we must move beyond an exclusive concern with individual buildings and explore the possibilities offered by integrated energy systems for communities.

COMMERCIALLY AVAILABLE DEVICES

Old consumption habits die hard, all the more so in a nation long accustomed to a ready supply of inexpensive energy. Nonetheless, the rude awakening of the oil embargo of 1973/74, which still conjures images of long lines and short supplies at the gas pump, prompted many of us to reconsider some fundamental attitudes and patterns of energy use. Since the embargo, the federal government and millions of Americans have responded to rising fuel prices (if not national calls to patriotism) by using commercially available devices as aids in taking simple but effective conservation measures. As increasing numbers of consumers have adopted these measures, a new set of values has begun to pervade the American consciousness: a set of values, motivations, and behavior based on the realization that conservation does not mean doing without, but doing more with the resources we have.

The federal government has undertaken the Federal Energy Management Program to increase energy efficiency in all federal operations, including its 400,000 buildings which consume approximately 45 percent of the energy used by the federal government. Between 1973 and 1975, the energy consumption of federal buildings was reduced about 20 percent, primarily as a result of operational changes such as lowering lighting levels, adjusting summer and winter thermostat settings, and installing improved controls on heating/cooling/ventilation (HVAC) systems. A large office building in Los Angeles reduced energy use 45 percent simply by modifying daily operational patterns—and providing direct feedback to users on the effectiveness of those modifications.

In the commercial sector, the federal government is encouraging the use of currently available but underutilized technologies through demonstrations in cooperation with building owners and manufacturers. Particular attention is focusing on schools and hospitals, which consume about two quads per year. The Office of Conservation and Solar Applications of the Department of Energy estimates that energy consumption can be reduced 30 percent in at least half the country's 100,000 schools and hospitals, which occupy almost seven billion square feet of space.

One area that offers substantial potential for saving energy in schools and hospitals is ventilation—the controlled entry of fresh air into building environments for health and safety purposes. Ventilation accounts for up to 75 percent of school heating requirements and 25 to 40 percent of hospital heating and air-conditioning requirements. Most existing ventilation standards were established without regard to energy efficiency, and many were established on no hard scientific basis.

A comprehensive federal effort is currently under way to establish a scientific basis for developing new ventilation standards that incorporate energy considerations for schools and hospitals. Selected hospitals are being retrofit with air-filtration systems, and preretrofit and postretrofit measurements of energy consumption and air quality are being taken. Energy conservation strategies for hospitals, including baseline ventilation requirements, will be recommended based on the results of these tests.

Many commercial establishments have taken the initiative to install automatic thermostat and lighting controls that enable them not only to save energy, but to take advantage of time-of-day electricity rates. Lighting controls are an example of how commercial buildings can improve energy efficiency without affecting daily operations.

A new ultrasonic system, for example, operates on the same principle as do ultrasonic burglar alarms, automatically turning lights on when a person enters a room and off after the person leaves. The amount of time that lights remain on can also be controlled with photocells which permit the use of daylight to supplement a building's electrically powered lighting system. Photocells are especially effective in controlling lights on the periphery of buildings and in buildings with a large proportion of window area (schools, for example). If a building's outer two or three rows of lights are connected to a photocell, those lights will turn off automatically on bright days without affecting the inner rows of lights. Building areas that do not require high lighting levels and have exposure to outside lighting (such as cafeterias and lobbies) are especially suited for the use of photocells.

Individuals, too, have been reaping immediate economic rewards by implementing changes in their own homes. Between two and three million residential attics, for example, have been insulated each year since the oil embargo. The simple addition of four to six inches of attic insulation reduces the amount of energy used for space heating by 10 to 20 percent, depending on the local climate and the condition of the house. Direct feedback in the form of lower energy bills not only rewards these homeowners, but motivates them to take additional energy conserving actions. On a national scale, the economic impact of a measure as simple as insulation takes on its proper perspective when one considers that space heating represents more than 50 percent of residential fuel use.

Automatic thermostat controls for residents offer another simple, inexpensive, and assured way of conserving energy. One such device, which can be attached easily to a conventional thermostat, sells for less than forty dollars in many hardware and department stores. In most regions of the country, a homeowner can save as much as 15 percent of the energy consumed for heating by using this device to reduce nighttime temperatures. In households occupied by a single working person or a working couple—both increasingly common—the device can also function during the day.

TECHNOLOGY IMPROVEMENT AND DEVELOPMENT

The immediate energy and dollar savings that can be derived from using simple devices and changing personal habits are impressive. The improvement of existing technologies also offers high savings in a short payback period (see Table 1), and the development and widespread adoption of new energy technologies offer even greater potential. The list of innovative energy technologies currently under development is virtually endless. Those being designed to increase energy efficiency in heating and cooling systems, which account for more than 60 percent of total residential energy use, are perhaps most important. Window materials and heat-pump systems are good examples of improved technologies that offer impressive energy saving potential.

Window Materials—One-fourth of the energy used to heat and cool our residential and commercial structures is wasted as a result of unwanted heat gains or losses through windows. The cost to the nation is about 5 percent of its annual energy consumption, or three quads, and

TABLE 1. SAVINGS ASSOCIATED WITH IMPROVEMENTS IN EXISTING TECHNOLOGY

Equipment	*Reduction in Annual Energy Use (%)*	*Payback Period (years)*
	Space Heating	
Gas furnace	25	6
Oil furnace	20	6
Heat pump	40	2-9
	Water Heating	
Electric	15	2
Gas	25	3
Refrigerator	50	2
Room air conditioner	35	6

Source: Eric Hirst and D. O'Neal, "Contributions of Improved Technologies to Reduced Residential Energy Growth," Paper No. 789374, *Proceedings of the 13th Intersociety Energy Conversion Engineering Conference,* August 1978.

the cost to a homeowner is $0.40 to $1.40 added to the annual fuel bill for every square foot of window area where a single-pane window is used.

This substantial loss of energy can be reduced easily and immediately through caulking and the installation of storm windows. Two energy conserving window materials are being developed and will soon be available commercially: heat mirrors and optical shutters. The two materials can be used either separately or together, depending on whether the climate requires space heating, cooling, or both. Heat mirrors are generally a worthwhile investment in northern climates, and optical shutters in southern climates.

Heat mirrors are thin film coatings with special optical properties that allow sunlight in to help heat a room but prevent internal heat from escaping. The energy savings that result when a heat mirror is laminated to a single-pane window are about equal to those derived from a conventional double-pane or storm window. However, because the materials cost less ($1.00 to $1.50 per square foot versus $1.00 to $2.00 per square foot for an interior storm window and $2.00 to $4.00 per square foot for an exterior storm window) and installation is easier, the payback in reduced fuel bills is more rapid. When applied directly to the inside of a storm window, a heat mirror nearly doubles the window's insulation value.

An optical shutter helps reduce unwanted heat gain in the summer by reducing the amount of sunlight entering a room, thus easing the air-conditioning load. One form of optical shutter is a gel-like substance sandwiched between two thin sheets of glass, which together form a single pane. Normally transparent, this substance automatically becomes translucent and reflective when the sunlight and room temperature exceed a certain level.

Heat Pump Systems—In recent years, as fuel prices have risen steadily, architects, engineers, builders, and government planners have discovered that the heat pump—originally developed in the 1930s—represents an attractive alternative to conventional electrical heating and cooling systems. In simplest terms, a heat pump is a device that pumps heat from a relatively cool area to another, warmer area. In its cooling mode, an air-to-air heat pump functions like an ordinary air conditioner by extracting heat from inside a building and pumping it outdoors. However, unlike an air conditioner, the heat pump can also work in reverse. During cold weather, it absorbs energy from the outdoor air and transfers it inside to heat a building.

In dramatic contrast to all of today's conventional systems, the heat pump delivers more heat than it consumes. Conventional systems use energy to *create* heat; heat pumps use energy to *transfer* and *intensify* heat already available in the surrounding environment. Because energy is needed only to run the fan and the compressor, and because already available heat is used, the heat pump can deliver as much as three times the energy it consumes.

"Advanced" heat pump systems can do even better. One of the most promising, the Annual Cycle Energy System (ACES), is currently being demonstrated by Oak Ridge National Laboratory and is scheduled for commercialization in the early 1980s. The federal government is currently supporting ACES installations in six residences and two commercial buildings. Impressed with the system's energy savings potential, several designers and builders have taken the initiative—without any federal financial support—to incorporate ACES into the design of twenty-one residences and eleven commercial buildings.

The principal components of ACES are a heat pump and an insulated tank of water which serves as an energy storage bin. During the winter, the heat pump draws heat from the water in the tank to heat the house and provide hot water. This extraction gradually turns the water into ice. In the summer months, the ice is used to provide air conditioning. The ice melts gradually, stores heat for the winter, and the cycle begins again.

The most detailed information available on the payback of ACES comes from a study of 406 apartments in sixty-eight buildings in Washington, D.C. These apartments are heated and air conditioned by a central boiler plant fired with natural gas. Twelve-month usage records indicate that 3,059 kilowatt-hours (kWh) of energy are required per apartment-year to provide heating, air conditioning, and hot water with ACES —a startling energy saving of 71 percent over a gas-fueled system. Compared to all-electric apartments, ACES could save 77 percent of the energy required, or 10,000 kWh per apartment-year. The system breaks even at any cost of energy over 2.5 cents per kWh.

The energy consumption of an ACES demonstration house is currently being compared with that of a control house (with an electric-resistance heating system) in Knoxville, Tennessee. Over the 1977/78 heating season, the ACES house consumed 5,500 kWh, and the control house consumed 14,880 kWh. Over the summer months, cooling was provided in the ACES house using stored ice until early August when the supply was exhausted and the heat pump had to be activated. Insulation was added to the floor of the ice-storage bin to reduce the heat leakage from the ground into the bin and thereby extend the use of ice further into the cooling season.

Experimental work is currently being conducted jointly by DOE and several utility and manufacturing companies to develop thermally activated heat pumps which would be powered by gas burned on-site rather than electricity generated off-site. Because the on-site use of gas is more cost-effective than the off-site generation of electricity, these new pumps should achieve an even greater return on investment than electric heat pumps. Furthermore, although these heat pumps will initially be designated to burn natural gas, future models will be able to use other energy sources, such as oil, coal, synthetic fuels, and solar energy.

INTEGRATED COMMUNITY ENERGY SYSTEMS

The energy efficiency of the residential and commercial sectors can be enhanced considerably through the design and implementation of integrated community energy systems (ICES) as well as the application of specific technologies. An ICES is not a new piece of hardware, but a comprehensive energy framework that represents an optimal combination of integrated technological systems, community designs, and implementation mechanisms.

Systems integration is achieved by combining the hardware of a community's energy systems into a single, unified whole capable of serving all the community's energy requirements. Traditionally, the suppliers of the various forms of energy used by a community have been single-purpose, competing utility companies; the result of this lack of coordination has been the loss of valuable energy. For example, a typical fossil fuel generating plant transfers only about 30 percent of the energy content of the fuel it burns to consumers in the form of electricity. The remaining 70 percent is lost as waste heat in the form of hot gases exhausted up a smoke stack and reject heat released to an adjacent body of water.

An ICES is designed to reduce this loss by exploiting the full potential of energy resources. Energy needed for the production of services is supplied in a series of steps; so that the energy requirements of one service are supplied using the by-products of the preceding step. For instance, when high-pressure, high-temperature steam is used to produce electricity, low-temperature steam or hot water is released as a by-product. This steam or hot water can then be used as the primary energy source for space heating and cooling. In turn, the low-temperature hot water that is a by-product of the space heating or cooling can be used for purposes such as snow melting.

Several federal projects are currently under way to develop and demonstrate ICES applications that are suitable for both new and existing communities. The grid-connected ICES uses a central, combined heat/power installation and distribution system to provide a community's heating, cooling, and electrical energy services. The basic elements of a grid-connected ICES are (1) a community energy plant, which operates primarily to supply thermal energy for heating and cooling nearby buildings and produces electricity as a by-product, and (2) a utility grid, which accepts excess electricity from the power plant when the community's electrical demand is low and supplies electricity when demand exceeds plant output. Alternative grid-connected ICES designs are adaptable to a variety of developments and are flexible in the types of fuels used.

Specific applications of grid-connected systems have deen designed and are currently being tested for economic feasibility at four sites: New Orleans, Louisiana; the University of Minnesota in Minneapolis; Clark University in Worcester, Massachusetts; and Trenton, New Jersey. Preliminary results at these four sites indicate that grid-connected ICES will save from 900 billion Btu's per year with a payback period of 2.9 years (in Louisiana) to 100 billion Btu's with a payback period of 10.9 years (in Massachusetts).

Technological integration in the central power plant enables a community to improve energy efficiency by extracting the most "work" out of its fuels. Even more important over the long term, such integration advances the economic feasibility of converting solar energy, urban waste, geothermal energy, or coal into usable forms of energy and thus paves the way for fuel switching.

But technological integration will yield these benefits only if supported by optimal community designs—carefully considered arrangements of buildings and facilities within a community. Although new communities have more opportunities for minimizing energy demand through physical layout, existing communities can also adopt a variety of strategies to reduce energy consumption, for example, by encouraging development near activity centers, channeling growth along existing transportation and utility lines, and incorporating energy considerations into land use decisions.

The federal government is exploring the integration of technological systems and community designs at two levels: small-scale site and neighborhood design, and energy master plans for entire communities. Energy efficient site and neighborhood design projects are currently being conducted for residential subdivisions, village centers, and new towns. In 1978, these projects focused on sites ranging from one hundred to five hundred acres in Radisson, New York; Greenbriar and Fairfax, Virginia; Shenandoah, Georgia; and Woodlands, Texas.

The ICES concept contains one more indispensable ingredient: implementation mechanisms to ensure an optimal combination of the integrated-systems and community-design elements. Communities of various sizes are serving as laboratories to test and demonstrate the integration of energy considerations into community-development planning. Preliminary plans were recently completed for several communities representing different development characteristics: a rural town (Mercer County, North Dakota), a new community (the Alaska state capital), and a redevelopment area (Atlantic City, New Jersey).

Harmonizing all these components will no doubt entail meeting some formidable challenges. Nonetheless, the sheer size of the market for the ICES concept—and the associated potential energy savings—suggests that the effort will be worthwhile. Urban renewal projects, new developments, entire downtown areas—the implications for reducing energy demand in the residential and commercial sectors merit careful consideration.

Barriers to Implementation

The history of serious national energy planning in this country is short, at best dating back four years. Our experience is sufficient, however, to confirm what some shrewd observers have suspected from the very outset—that the nontechnological problems would prove to be the most challenging of all. We have entered the post-1973 period endowed with a rich legacy of scientific and engineering talent. The technological base for improving energy efficiency in our buildings is not only well-established, but also poised for further advances. Yet, it is now apparent that the technology of energy efficiency is not easily transferred to the day-to-day world and that the road to conservation is marked with formidable behavioral, institutional, and legal/regulatory barriers.

BEHAVIORAL BARRIERS

The transition we must make from profligate to efficient energy use must take place as much—in fact, first—in our minds as in our laboratories and our buildings. At a minimum, this mental transition will require national awareness of the true dimensions of our energy situation.

Perhaps too often the energy crisis is depicted in bewildering graphs, imposing tables, and arcane projections. Although these are unquestionably essential tools in energy planning, the most effective way of first informing—and then motivating—the general public to adopt energy saving measures is to present practical information straightforwardly and directly. At the most fundamental level, consumers must be made aware of the energy cost implications of buying a home, heating an apartment, or purchasing an appliance. Educating consumers about one central concept, life cycle costing, will go a long way toward creating the kind of awareness that compels action.

There is certainly nothing esoteric about the concept of energy life cycle costing. Although it may take on a certain sophistication in the hands of a professional accountant or economist, its beauty lies in that it can be easily understood and readily applied. A good working definition of life cycle costing might run as follows: in today's—and tomorrow's—energy market, a low *initial* price for a home or an appliance does not necessarily mean a low *overall* cost. In making a purchase decision, the consumer must therefore consider not only the price of buying the home or appliance, but the energy cost of operating it as well. For example, the initial cost of a truly energy efficient home may exceed the cost of an ordinary home by, say, $2,500. But when the cost of fueling the ordinary home over a number of years is taken into account, the overall, or life cycle, cost of the energy efficient home turns out to be less than that of the conventional home. Moreover, because it contains energy saving features (solar heating, a heat pump, or complete insulation, for example), the resale value of the efficient home will exceed that of the ordinary home.

Traditionally, most individuals—and most institutions for that matter—have made purchase decisions almost exclusively on the basis of initial cost. As more and more consumers come to realize that the upward trend in energy prices is not a passing phase but a long-term reality, the principles of life cycle costing are likely to become the foundations of energy related decisions in all sectors of the economy. This change in values and motivation will occur gradually, however, as part of the long-term consumer response to rising energy prices discussed by Pindyck in Chapter 2.

INSTITUTIONAL BARRIERS

Behavioral barriers are common to many sectors. The buildings sector, however, presents some unique and especially difficult institutional barriers, largely as a result of its fragmentation, diversity, and inherent slowness to change.

The fragmented nature of the sector is readily apparent. The nation's stock of residential buildings alone numbers approximately 74 million units. The design, construction, financing, and operation of these buildings involve a multitude of participants, including millions of homeowners and renters; countless developers, architects, engineers, labor unions, and contractors; and numerous lawyers, financial institutions, and insurance firms. If we set out to improve energy efficiency in the automobile industry, a great deal can be accomplished by working directly with a few manufacturers in Detroit. But where, and how, do we embark upon the task of transforming our buildings into energy efficient structures? Even the logical first step of compiling and disseminating useful, consistent information to the many participants in the building process poses formidable challenges.

These participants play out their various roles in diverse environments and under differing conditions. Again, to take the automobile as a point of comparison, a car driven in New England will differ from a car driven in Arizona only in that the former will require snow tires and antifreeze in the winter. But the climatic and regional differences that prevail throughout our vast country have brought into being a wide variety of building types. Indeed, if we do our job well—that is, if we design buildings that accommodate rather than confront their immediate environments—the diversity of building types will increase in the future.

Another factor impeding the timely modification of energy consumption habits in the buildings sector is the sector's inherent slowness to change. Sixty percent of our current stock of buildings will still be operational in the year 2000. Any kind of change must therefore evolve first through the retrofit of existing buildings and then through the gradual incorporation of energy efficient systems into new building designs.

LEGAL/REGULATORY BARRIERS

Differences in climates, fuel availability and prices, and building types no doubt pose challenges to attaining energy efficiency in the buildings sector. Legal/regulatory variations, however, may well impede progress. Each state, each county, each community has its own building codes, zoning ordinances, utility regulations, taxing authorities, and management structures. Originally established to ensure compliance with fire, health, and safety requirements, building codes, for example, have only recently begun to reflect energy concerns.

Introducing energy efficiency standards into our building codes is a complex task. Sensitive questions about local and state jurisdictions will have to be resolved, thousands of codes will have to be revised, and even more architects, builders, and manufacturers will have to be educated about energy efficiency.

Another difficult question arises with regard to building ownership. If the occupant of a commercial or residential building is not the owner, where is the incentive to introduce energy saving modifications? Perhaps even more important, if utility charges are included in the rent payment or if utility services are not individually metered, where is the incentive even to turn down thermostats? Uncertainty over future fuel prices and the cost-effectiveness of different conservation measures will strongly influence the attitudes of both building owners and occupants.

Evidently, if we are to improve energy efficiency in our residential and commercial buildings, we must address legal/regulatory issues at the same time that we work to modify behavioral patterns and overcome institutional barriers.

Possible Federal Roles

Federal, state, and local governments; the buildings industry; utilities and regulatory commissions; banks and savings and loan associations; and, most importantly, the public at large—each must play a role if we are to surmount the barriers and improve energy efficiency in the buildings sector. On a small scale, voluntary conservation measures can yield immediate results. On a broader scale, the ultimate success or failure of any energy technology or system will be determined in the commercial arena. Clearly, the private sector will play a crucial role in making our buildings energy efficient. Nevertheless, it appears that the federal government must assume a lead role, considering the large capital outlays and economic risks involved in energy related RD&D, the formidable nontechnological barriers that can slow improved efficiency in the residential and commercial sectors, and the importance of our energy future to national interests.

In the National Energy Plan, President Carter set forth a number of broad objectives for energy conservation in residential and commercial buildings, including:

1. Bring 90 percent of existing homes up to minimum energy efficiency standards,
2. Accelerate the adoption of minimum energy efficiency standards for all new buildings,
3. Encourage the weatherization (retrofit) of existing buildings to reduce energy losses and improve the efficiency of energy using systems,
4. Assign priority to conserving energy in hospitals and schools,
5. Reduce energy use in existing federal buildings 20 percent by 1985, and in new buildings 45 percent by 1985 over 1975 levels, and
6. Develop mandatory minimum energy standards for major appliances such as furnaces, water heaters, and refrigerators.

Federal leadership in achieving these objectives can take a variety of forms. Five generic types of action together provide an overview of current and planned federal initiatives: (1) sponsorship of RD&D, (2) regulation, (3) financial incentives, (4) coordination of disparate but ideally combined activities (e.g., building codes), and (5) dissemination of information about energy efficiency technologies and measures.

DOE, the primary federal agency charged with developing energy programs and technologies, has two overriding RD&D objectives: (1) to accelerate the development of technologies that would not be developed as quickly—or at all—because of the technical risk and economic uncertainty involved, and (2) to ensure that the RD&D essential to the development and application of the most cost-effective energy saving options is performed. In general, corporations focus their resources on developing technologies that are in their best short-term interests; they are not concerned with researching and developing technologies needed to solve national problems. Nonetheless, firms involved in technology research and development (R&D) have often looked to the government for funding and then for post-R&D actions to ensure the commercial viability of the technologies they develop.

They have not looked in vain. Government funding for energy related RD&D has grown dramatically in recent years. In 1977, DOE's predecessor, the Energy Research and Development Administration, surpassed the National Aeronautics and Space Administration to become the government's second largest R&D contractor. (The Department of Defense remains the first.) Although conservation represents a small portion of the government's energy budget, its allotment has grown considerably in recent years: from $12 million to $55 million for conservation in buildings and community systems from 1975 to 1978.

This growing commitment to energy RD&D has been supported by both legislative measures and financial incentives aimed at fostering energy conservation. These initiatives stand as vivid testimony to how much can be accomplished outside our research laboratories.

The same holds true for the government's role as a coordinator of disparate activities and as a disseminator of practical information about energy technologies and planning tools. A good example of how the federal government is helping to accelerate consumer response to national energy priorities is through regulatory actions in the complex and hitherto confused area of building standards and codes.

Between 1974 and 1976, the federal government enacted several pieces of legislation that included requirements for energy performance standards in buildings. Under the Energy Conservation Standards for New Buildings Act of 1976, DOE and the Department of Housing and Urban Development are developing energy performance standards for new residential and commercial buildings. To guide state and local governments in adopting and implementing these standards, DOE is developing energy performance codes for various building types in various climatic regions of the country. These codes are being developed in conjunction with representatives of the officials who will eventually be responsible for enforcing them—a good example of how federal support need not mean undue federal interference. DOE is working directly with groups such as the National Conference of States on Building Codes and Standards; the American Society of Heating, Refrigerating and Air Conditioning Engineers; and the American Institute of Architects.

An example of the federal government acting as a catalyst and information provider for state and local actions is the Energy Extension Service. A pilot program is currently under way in ten states; each state is designing and operating its own program to provide direct technical assistance and practical information on energy conservation technologies and techniques to homeowners, small businesses, and the groups that influence their energy consumption (e.g., builders, architects, financial institutions). A major thrust of the Energy Extension Service is to build upon ongoing programs and existing information networks in each state (e.g., homebuilders' associations, local chambers of commerce) and thereby create a credible, central source of direct assistance and information instead of merely establishing another level of bureaucracy.

The government is especially well-equipped to gather and coordinate information from a wide variety of sources and then convey that information to individuals and communities. The federal role in helping communities integrate energy considerations into their planning is a good example. Although many energy conservation planning tools are available, municipalities cannot usually afford to have their urban planners and city managers develop and adapt those tools to meet their particular needs.

The government recently began the first of some two-year testing and demonstration programs in a number of local municipalities and areawide agencies. The results will be disseminated through the National Governors Association, the League of Cities, the Council of Mayors, and other interested associations. This effort will serve to enhance local understanding of the relationships among patterns of community development, land uses, and levels of energy consumption and thereby help to create precisely the kind of energy efficient communities we so badly need.

The Implications of Federal Actions

There is no doubt that federal actions can reduce the growth of energy consumption in the residential and commercial sectors and thus limit the amount of money that households and businesses must spend to meet their energy needs. It is equally clear that the resultant savings in energy and dollars will far exceed costs on a national level. The question is: under what set of conditions and through what actions will we reap the richest harvest?

Projecting future trends is both an art and a science. The nation's energy future will be influenced by a variety of factors, including: the types of energy saving equipment available, how much of that equipment consumers purchase, fuel prices, and the pace of economic growth. In the nontechnological arena, the impact of a national energy policy will hinge on the types of regulations adopted, how strictly they are construed, the extent to which the public is willing to accept them, and how successfully they are enforced.

With or without federal action, energy use in the residential and commercial sectors is projected to grow at a slower rate over the remainder of this century than it has in the past, as a result of consumer response to projected fuel price increases. If the government takes current authorized and proposed measures to foster energy conservation, average annual growth in energy use is expected to decline to less than 1 percent in the residential sector and less than 3 percent in the commercial sector, compared with historical (1950–1972) growth rates of 4 percent and 4.7 percent, respectively.

Federal efforts will yield substantial energy and economic benefits to the nation, as illustrated for the residential sector in Figure 6. The savings associated with regulatory/incentive programs will start out larger and take effect sooner than those associated with RD&D efforts, but then will grow more slowly. The savings associated with new energy technologies will begin more slowly but accelerate over time as more new systems are installed and as fuel prices continue to rise.

The separate and combined impacts of several specific government actions can be projected to the year 2000 in terms of three scenarios. Scenario I assumes the status quo, that is, no change in real fuel prices. Although this theoretical scenario is generally considered unrealistic, it provides a yardstick for measuring the effects of other independent changes.

Scenario II postulates that fuel prices rise substantially but that no federal conservation efforts are undertaken. This "baseline" projection is used to measure voluntary consumer responses to rising prices. Scenario III assumes the same rise in fuel prices, but adds two other ingredients:

Fig. 6. Residential Energy Savings Associated with Regulatory/Incentive and RD&D Programs

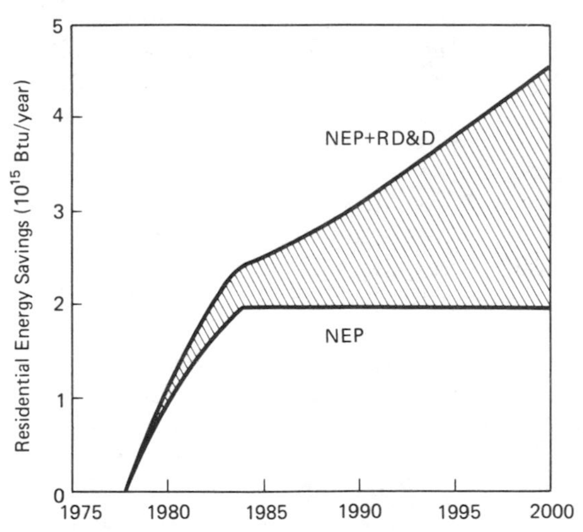

Source: Eric Hirst, *Energy and Economic Benefits of Residential Energy Conservation RD&D,* Oak Ridge National Laboratory, ORNL/CON-22, February 1978.

federal regulatory/incentive programs and federal and private RD&D programs.

The methodology for predicting the outcome in each of these scenarios draws upon detailed engineering/economic models of residential and commercial energy use developed by Oak Ridge National Laboratory. The models simulate household and commercial energy use at the national level for 296 fuel/end use combinations. Each of these combinations is evaluated for each year of the simulation as functions of: stocks of occupied housing units and commercial floor space, plus new construction; equipment ownership by fuel and end use; thermal integrity of structures; average unit energy requirements for each type of equipment; and usage factors that reflect occupant behavior. The models also calculate annual fuel expenditures, equipment costs, and capital costs for improving the thermal integrity of new and existing structures. These cost figures provide the basis for developing simple benefit/cost measures for each program evaluated.

The data used in the models include population, fuel prices, per capita income, and specifications for federal conservation programs (e.g., efficiency standards, tax incentives for retrofitting homes, development of new technologies). Each of these inputs is provided for the 1970–2000 period.

SCENARIO I (STATUS QUO PROJECTION)

The first scenario assumes that: (1) real fuel prices remain at 1976 levels until the year 2000, (2) no federal programs are undertaken to encourage reduced energy consumption, and (3) consumers continue making changes begun in the early 1970s, such as retrofitting homes with energy saving materials and devices.

Under these assumptions, energy use in buildings will grow at an average annual rate of 3.4 percent between 1977 and 2000, compared with the significantly higher rate of 4.3 percent during the 1950–1972 period. Residential energy use will grow more slowly, at 2.2 percent per year, while commercial energy use will grow more rapidly, at 4.8 percent per year, primarily because commercial activity is expected to grow at a much faster rate than overall GNP. Cumulative energy use in the buildings sector between 1977 and 2000 is expected to total 942 quads.

Energy consumption is not projected to grow at as fast a rate as it has in the past for several reasons. First, real fuel prices will remain constant; historically, prices have declined, encouraging consumers to use increasing amounts of energy. Second, the slower rate of population growth will mean slower growth in households and commercial floor space. Finally, the possible applications of energy use to buildings functions, which expanded dramatically in the past, have been nearly exhausted.

SCENARIO II (BASELINE PROJECTION)

Scenario II rests upon conditions identical to Scenario I, with one major exception: real fuel prices are assumed to increase over time at rates forecast by DOE. Up to the year 2000, substantial increases are anticipated in real gas prices (average of 3.1 percent per year) and moderate increases are anticipated in electricity (1 percent) and oil (1.7 percent) prices.

Under these conditions, the growth in annual household energy consumption will be much slower than it has been in the past—1.8 percent, down from 4 percent from 1950 to 1972—as a result of voluntary consumer reactions to rising fuel bills. A similar, but smaller, reduction is projected for the commercial sector, where growth in average annual energy use is expected to decline to 3.9 percent, down from the historical rate of 4.7 percent. The cumulative total of energy used for residential and commercial needs is projected at 858 quads, which is 9 percent less than if real fuel prices were to remain constant. These reductions in residential and commercial energy use caused by rising fuel prices are consistent with the aggregate own-price elasticities presented by Pindyck in Chapter 2.

SCENARIO III (GOVERNMENT REGULATION AND RD&D)

The third scenario assumes that fuel prices rise as estimated by DOE in the baseline projection, but introduces a major change: that the federal government undertakes energy conservation regulatory/incentive programs and fosters federal and private RD&D in both the residential and commercial sectors. The differences between Scenarios II and III (see Tables 2 and 3) clearly show the value of federal programs to encourage energy efficiency in the buildings sector. Energy use will be reduced almost 20 percent (nine quads) in the year 2000 as a result of these programs. The overall rate of energy growth between 1977 and 2000 will slow from 2.7 to 1.8 percent per year. The net economic benefit to the buildings sector of the federal regulatory/incentive and RD&D programs will be almost $50 billion.

Residential Sector—In the residential sector, the federal regulatory actions assumed in Scenario III are those conservation programs authorized by Congress in the Energy Policy and Conservation Act of 1975 and the Energy Conservation and Production Act of 1976 and expanded in the President's National Energy Plan (see Table 4). The effect of these programs to achieve appliance efficiency targets, implement thermal standards for the construction of new buildings by 1980, and encourage households to weatherize (retrofit) existing structures, will be to reduce residential energy growth from 1.8 percent per year (Scenario II) to 1.4

TABLE 2. RESIDENTIAL ENERGY USE PROJECTIONS

Scenario	Description	*Energy Use* Cumulative 1977–2000 (QBtu)	Average Annual Growth Rate 1976–2000 (%)
I.	Constant real fuel prices, no federal programs	533	2.2
II.	Baseline: fuel prices rise according to DOE projections, no federal programs	502	1.8
III.	DOE fuel-price estimates		
	–federal regulation/ incentives	460	1.4
	–federal regulation/ incentives and RD&D	438	0.9

Source: Eric Hirst and Janet Carney, *Residential Energy Use to the Year 2000: Conservation and Economics,* Oak Ridge National Laboratory, ORNL/CON-13, September 1977; and Eric Hirst, *Energy and Economic Benefits of Residential Energy Conservation RD&D,* Oak Ridge National Laboratory, ORNL/CON-22, February 1978.

TABLE 3. COMMERCIAL ENERGY USE PROJECTIONS

Scenario	Description	*Energy Use* Cumulative 1978–2000 (QBtu)	Average Annual Growth Rate 1977–2000 (%)
I.	Constant real fuel prices, no federal programs	409	4.8
II.	Baseline: fuel prices rise according to DOE projections, no federal programs	356	3.9
III.	DOE fuel-price estimates		
	–federal regulation/ incentives	314	3.1
	–federal regulation/ incentives and RD&D	307	2.8

Source: Jerry R. Jackson, *Energy Use and Conservation in the Commercial Sector: An Econometric-Engineering Analysis,* Oak Ridge National Laboratory, ORNL/CON-30, unpublished draft.

TABLE 4. RECENT FEDERAL LEGISLATION AND PROPOSALS AFFECTING RESIDENTIAL ENERGY USE

Energy Policy and Conservation Act (PL 94–163, December 22, 1975)
Residential equipment and appliance labeling
Residential equipment and appliance efficiency targets
State energy conservation plans

Energy Conservation and Production Act (PL 94–385, August 14, 1976)
Thermal standards for new buildings
Financial assistance to weatherize existing buildings
State conservation plans:
- Public education programs
- Energy audits

Conservation assistance for existing buildings
- Demonstration programs
- Financial assistance

Energy conservation obligation guarantees

National Energy Plan (April 20, 1977)
Efficiency standards for new buildings
Conservation program for existing buildings
- Tax credits to homeowners
- Utility conservation programs
- Provision of capital at low interest rates
- Increased funding for low-income weatherization program
- Rural home-weatherization program
- Tax credits to commercial businesses
- Grants to schools and hospitals

Reduced energy use in existing and new federal buildings
Appliance-efficiency standards

Source: Eric Hirst and Janet Carney, *Residential Energy Use to the Year 2000: Conservation and Economics,* Oak Ridge National Laboratory, ORNL/CON-13, September 1977.

percent per year. The cumulative energy saving associated with these programs is estimated at forty-two quads.

Even though portions of the National Energy Plan will increase residential fuel prices (e.g., crude oil equalization tax, changes in natural gas price regulation), efficiency improvements to appliances and buildings will be more than offset by these increases. The net present worth of household energy related dollar benefits will amount to almost $2 billion.

The potential energy and dollar savings expected to accrue as a result of federal and private RD&D programs are comparable to those associated with government regulatory/incentive programs. If RD&D programs to improve efficiency in residential structures, heating and air-conditioning equipment, and appliances and lighting accelerate technology commercialization as projected, the cumulative energy saving will be twenty-two quads, which is about 5 percent less than projected energy consumption if the government sponsors regulatory/incentive programs only. By the year 2000, the *annual* energy saving will be 2.4 quads, 10 percent less than if no RD&D is sponsored. The net economic benefit to households will be $21 billion.

Commercial Sector—The effect of regulatory/incentive programs to implement thermal standards for the construction of new buildings, provide tax credits and grants for the retrofit of existing buildings, and reduce energy use in federal buildings through the Federal Energy Management Program will be to reduce commercial energy use forty-two quads, a 12 percent reduction, between 1977 and 2000. Commercial energy growth will slow from 3.9 to 3.1 percent per year. The net economic benefit to the commercial sector of these programs is estimated to be $20 billion.

Developing new technologies for the design, construction, and operation of commercial buildings and HVAC equipment is expected to save even more energy: an additional seven quads by 2000. The costs associated with this RD&D will be more than offset by fuel bill reductions, for a net benefit of almost $6 billion.

Conclusions

With or without federal action, energy use in the residential and commercial sectors is projected to grow at a slower rate over the remainder of this century than it has in the past as a result of consumer response to projected fuel price increases (as Pindyck suggests in his investigation of the price elasticity of energy). Price increases alone will not be sufficient, however, to motivate most consumers to modify their energy consumption patterns.

Federal action will yield substantial energy and economic benefits to the nation under any scenario. The savings associated with regulatory/incentive programs are expected to start out larger and take effect sooner than those associated with RD&D efforts, but then grow more slowly. The savings associated with new energy technologies will begin more slowly but accelerate over time as more new systems are installed and as fuel prices continue to rise. An effective energy policy should thus emphasize regulatory/incentive programs and RD&D concurrently, so that savings from short-term programs are enhanced as long-term projects come to maturity.

We must also remain alert to the possibilities of fruitful interaction between regulatory/incentive and RD&D programs. Appliance and construction standards, for example, may spur new RD&D projects, and federally sponsored RD&D may lead to more effective regulatory standards. In such interactive ways, the total effect of a combined federal program may well prove to be greater than the sum of its parts.

The federal government cannot work alone, however. Unless the multitude of participants in the residential and commercial buildings sectors—developers, architects, builders, labor unions, financial institutions, state and local government agencies, and homeowners—all play an active role, certain barriers to achieving energy efficiency will never be surmounted. Only by exploiting existing technologies while developing new technologies; by confronting behavioral, institutional, and legal/regulatory barriers simultaneously; and by combining federal and private forces will the nation achieve the level of energy efficiency that is essential to its economic, political, and social well-being.

Robert O. Reid and
Melvin H. Chiogioji

6

Technological Options for Improving Energy Efficiency in Industry and Agriculture

Introduction

The industrial sector is the single largest user of energy in the United States. It accounts for roughly 40 percent of final end use energy requirements. In addition to being the largest user of all energy forms, it is the largest consumer of natural gas and second only to personal passenger vehicles in its consumption of petroleum products. Although the percentage varies, roughly two-thirds of the industrial sector's energy requirements are satisfied by oil or natural gas.

Difficulties encountered in attempting to forecast future energy consumption practices in the industrial sector stem from several diverse but related factors. First, the industrial sector cannot be viewed as a monolithic structure which responds to any single set of stimuli. Composed of firms engaged in construction, mining, agriculture, and manufacturing, the industrial sector is the most heterogeneous of the three end use sectors of the economy. Energy is used for every conceivable purpose and in

---

ROBERT O. REID *is Executive Vice President of Energy and Environmental Analysis, Inc., specializing in regulatory economics and environmental impact analysis. In 1972–74 Dr. Reid was Assistant Director of the Office of Policy Analysis of the Environmental Protection Agency.*

MELVIN H. CHIOGIOJI *is Assistant Director for Systems Analysis in the U.S. Department of Energy.*

almost every conceivable form. Uses range from combustion in boilers to highly complex chemical and catalytic reactions which produce some of our most sophisticated and important petrochemicals.

A second problem when addressing energy consumption in the industrial sector is the integrity of the production process. While energy is an important factor of production, it cannot be separated from the system within which it is being used. This issue relates to the concept of energy productivity. Whereas it may be technically feasible to improve the ratio of product output to energy input, realization of these improvements will depend totally on whether the integrity of the production process can be maintained.

The third question is one of opportunity to optimize energy requirements in the face of higher prices. For years the U.S. industrial sector faced declining real energy prices. Promotional prices for natural gas and electricity were the rule rather than the exception. Secure supplies at low cost were assumed. This situation no longer exists, and the industrial sector is struggling with the situation. While much has been done in the near term to reduce fuel consumption through improved energy consumption practices, available technological options have been known for decades. Whether efforts to improve energy productivity will continue in the future will depend on the success of research and the strength of the American economy.

The question of energy productivity in the industrial sector cannot be discussed in the abstract. Technical engineering and economic principals must be the guide to informed public policy as it affects this vital portion of the American economy. The industrial sector is faced on the one hand with probably the most inefficient (in a technical sense) capital base of any industrialized nation in the world. In the past, energy was cheap, therefore it was exploited liberally in an attempt to produce less expensive goods and services. Unless we are to reduce our standard of living substantially, we will have to continue to rely on plants and equipment which were built during periods of low energy prices and which, therefore, are less efficient than required by current engineering practices.

While facing severe constraints with regard to existing capital stock, as new plants and equipment are brought on-line, industries will have significant opportunities to regain a dominant position. U.S. industry does not have to be any less efficient in the long run than other industrialized nations. The question is whether the marketplace will provide the proper prices and research activities will be adequately funded. In the long run, with its advanced technological and scientific skills and the prospect of an enhanced role for its abundant coal supplies, the U.S. could simultaneously rationalize its use of scarce energy supplies while continuing to have the lowest average overall cost of energy resources.

Fundamentals of Improving Energy Productivity

The most important point to recognize is that energy cannot be separated from the production process. Regardless of whether it is being used to produce pharmaceuticals, drive a drill press, or power a combine, energy is integral to the production process. Energy efficiency, therefore, must be viewed within the boundaries of technical and economic feasibility, neither of which can be considered static concepts.

Technical options for improving the energy use characteristics of the industrial sector can be grouped into four reasonably discrete categories.

1. *Housekeeping.* Actions in this category are similar to those taken by the average consumer attempting to reduce his heating bill. The list of actions which can be taken, however, is considerably larger. They range from reducing thermostat settings to optimizing air to fuel ratios in existing combustion facilities.
2. *Waste heat recovery.* Although most of the industrial sector's capital stock is fixed, a broad range of technologies can be retrofit into existing plants to improve the overall energy efficiency of the process. Actions here range from improvements in heat exchanger systems to installation of expansion turbines on existing fluid catalytic cracking units in petroleum refineries to produce electricity from a previously wasted source of energy.
3. *Process changes.* Opportunities in this category begin to affect directly the integrity of the production process. Any process change leads to fundamental alteration in the production process, which permits greater use of land, labor, or capital at a reduction in the need for energy. Although the distinction is not totally black and white, process changes generally are only feasible in the context of new plant investments. Examples include continuous casting in the steel industry or direct reduction of aluminum.
4. *Production changes.* The final category is one that is exceedingly difficult to quantify but could represent the largest potential opportunity for energy savings. Changes range from restructuring product lines to satisfy existing demands with a less energy intensive substitute product to altering the basic design of a product so that it consumes less energy when placed in service. Examples include using slag as a raw material to produce cement and producing more efficient automobiles for use by the general public.

The opportunities to improve energy productivity in the industrial and agricultural sectors represented by these four categories are too numerous to list, but a few examples are helpful as illustrations.

HOUSEKEEPING

Every plant manager has his or her story of how large energy savings resulted from a simple change in standard operating procedures.

Whether or not these changes are improvements in energy productivity or simply elimination of waste is questionable. But regardless of how they are characterized in the aggregate, improved housekeeping can result in billions of dollars of savings at little or no real cost.

Housekeeping generally is categorized as any action which does not involve capital expenditures. Thus, such measures generally involve operating and maintenance procedures. They are generally labor intensive, which can be translated directly into more jobs.

Simple examples are helpful in illustrating housekeeping improvements. While estimates vary, at a minimum, five to six quadrillion Btu's are used each year to produce steam in industrial plants. The steam is cascaded through the production process and normally returns to the boiler as condensate. Inspection and maintenance of a steam system is important. Initial inspections commonly reveal that as high as 7 percent of the steam traps in any one system leak. At high pressures, even a relatively small hole can lead to large steam losses. For example, a hole 0.1 inch in diameter on a 200 pounds per square inch gauge (psig) steam line can lead to an annual heat loss of 750 MMBtu's per year. With fuel prices in the range of $1.50 per MMBtu, the annual loss in this example would be $1,125. Such a loss could result from a trap simply being stuck open, and the solution would be a monthly or bimonthly inspection.

Such examples are not difficult to develop. Most industry trade associations and several studies performed by the federal government have documented these opportunities. The savings potentials are enormous and represent one of the principal means by which existing industrial plants can reduce their energy requirements while maintaining production schedules.

WASTE HEAT RECOVERY

Waste heat normally is defined as heat rejected from a process at temperatures above ambient conditions. The value of this waste heat is a function of the cost of recovering additional heating value from it and the opportunity cost of an equivalent amount of fuel. Waste heat is categorized into three temperature ranges: high (above 1200° F), medium (400° F to 1200° F), and low (below 400° F). In general, the higher the temperature the greater the potential value, although the quality of the medium containing the waste heat and the quantity available are also important considerations.

Several surveys of waste heat sources have been compiled. High temperature waste heat commonly is associated with direct firing in process furnaces, where temperatures reach up to 3000° F. Medium temperature waste gas sources include boiler and gas turbine exhausts, as well as several other generic processes. Low temperature waste gas sources are found

TABLE 1. SOURCES OF WASTE HEAT BY TEMPERATURE RANGE

High Temperature Waste Gas Sources		Medium Temperature Waste Gas Sources		Low Temperature Waste Gas Sources	
Type of Device	*Temperature (°F.)*	*Type of Device*	*Temperature (°F.)*	*Source*	*Temperature (°F.)*
Nickel refining furnace	2500–3000	Steam boiler exhausts	450–900	Process steam condensate	130–190
Aluminum refining furnace	1200–1400	Gas turbine exhausts	700–1000	Cooling water from:	
Zinc refining furnace	1400–2000	Reciprocating engine exhausts	600–1100	• Furnace doors	90–130
Copper refining furnace	1400–1500	Reciprocating engine exhausts (turbo charged)	450–700	• Bearings	90–190
Steel heating furnaces	1700–1900	Heat treating furnaces	800–1200	• Welding machines	90–190
Copper reverberatory furnaces	1650–2000	Drying and baking ovens	450–1100	• Injection molding machines	90–190
Open hearth furnace	1200–1300	Catalytic crackers	800–1200	• Annealing furnaces	150–450
Cement kiln (Dry process)	1150–1350	Annealing furnace cooling systems	800–1200	• Forming dies	80–190
Glass melting furnace	1800–2800			• Air compressors	80–120
Hydrogen plants	1200–1800			• Pumps	80–190
Solid waste incinerators	1200–1800			• Internal combustion engines	150–250
Fume incinerators	1200–2600			Air conditioning and refrigeration condensers	90–110
				Liquid still condensers	90–190
				Drying, baking, and curing ovens	200–450
				Hot processed liquids	90–450
				Hot processed solids	200–450

Source: W. M. Rohrer and K. Kreider, "Sources and Uses of Waste Heat," *Waste Heat Management Guidebook,* NBS Handbook 121 (Washington, D.C.: Government Printing Office, January 1977), p. 5.

throughout the industrial sector. Table 1 shows common sources for all three categories.

Methods of recovering waste heat generally can be classified into the following categories:

1. *Direct utilization,* e.g., for drying or preheating process material when no external heat exchanger is employed.
2. *Recuperation,* in which waste gases and air or other gas for preheating are separated by a metallic (or in cases of very high temperatures, a refractory) heat exchange surface. Energy is transferred from one gas to another continuously.
3. *Regeneration,* in which heat from waste gas is conducted to and stored in a heat exchange medium, such as a refractory, and subsequently heats air for preheating.
4. *Waste heat boiler,* a form of recuperation in which hot waste gases generate process steam or hot water.

To use the waste heat from the sources listed in Table 1, heat must be transferred from the waste stream to the process stream. Waste heat may be used to preheat combustion air, boiler feedwater, and liquid or solid feedstock. If temperatures are sufficiently high, waste heat can be used to generate steam for the production of electrical power, mechanical power, process steam, or any combination of the above.

Table 2 summarizes the types of systems available to recover waste fuel up to 1000° F from various sources. In general, there is some form of heat recovery system available for every waste heat source. However, the availability of a system does not necessarily determine its use. The economics of waste heat recovery will depend on: (1) whether there is an adequate use for the waste heat within the plant, (2) whether an adequate quantity is available to achieve economies of scale, (3) whether the quality of the waste heat is consistent with its intended end use, and (4) the distance over which the waste heat must be transported. All of these factors will determine whether the waste heat can be recovered at a cost which promises a sufficiently high return on invested capital. If these factors are to be satisfied, it is still important to consider the installation cost and the value of the Btu's recovered.

Figure 1 shows a typical display of the cost of recovering waste heat in various temperature ranges as a function of the quantity of waste heat recovered. This particular application is a gas to gas heat exchanger. For a 500° F waste heat source, the cost of recovering four MMBtu's per hour can range from $120,000 to $200,000. If the system operates half the year, the recovered energy valued at $1.50 per MMBtu would be roughly $26,000. The payback for this system therefore would be in the range of 4.6 to 7.7 years, which generally is inadequate to justify the investment.

TABLE 2. FEASIBILITY OF THERMAL WASTE ENERGY RECOVERY FROM IDENTIFIED INDUSTRIAL SOURCES (STATE-OF-THE-ART TECHNOLOGY UNLESS NOTED)

Technologies	*Cooling Water at 90–130 °F.*	*Condensate at 140–200 °F.*	*Process Water at 100–140 °F.*	*Boiler Exhaust at 300–600 °F.*	*Furnace Exhaust at 500–1000 °F.*
Power systems					
• Rankine	No	Yes	No	Yes	Yes
• Sterling	No	No	No	Yes *	Yes *
Heating systems					
• Heat exchangers					
–Liquid/liquid	Yes	Yes	Yes	N/A	N/A
–Liquid/gas	Yes	Yes	Yes	N/A	N/A
–Gas/liquid	N/A	N/A	N/A	Yes	Yes
–Gas/gas	N/A	N/A	N/A	Yes	Yes
–Gas/boiling	N/A	N/A	N/A	Yes	Yes
• Heat pumps (external power required)	Yes	Yes	Yes	N/A	N/A
Cooling systems					
• Absorption	No	Yes	No	Yes	Yes
• Rankine-cycle driven	No	Yes	No	Yes	Yes
• Steam jet	No	No	No	Yes	Yes

Source: Drexel University et al., *Industrial Applications Study,* Volume III, "Technology Data Base Evaluation of Waste Recovery Systems," prepared for the Energy Research and Development Administration, January 1977.
N/A–Application Not Appropriate.
No–Temperature insufficient even in near term.
Yes–Temperature sufficient, state-of-the-art technology.
Yes *–Temperature sufficient, near-term technology.

Fig. 1. Estimated Heat Exchanger Installed Costs

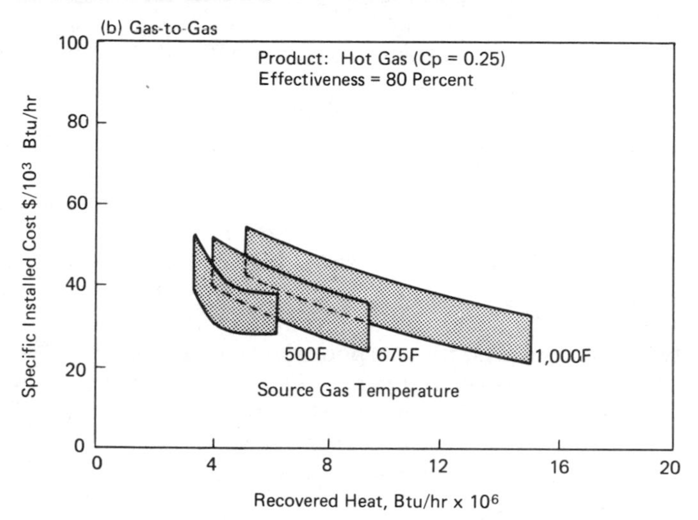

Band reflects variations in gas-side pressure drop.
Source gas exit temperature = 300° F.
Source: Drexel University et al., *Industrial Applications Study,* Volume III, "Technology Data Base Evaluation of Waste Recovery Systems," prepared for the Energy Research and Development Administration, January 1977.

However, with a waste heat source capable of delivering twelve MMBtu's per hour, the cost would range from $260,000 to $450,000. The value of the recovered energy in this instance would be roughly $80,000, which would provide for a payback in the range of 3.25 to 5.6 years. Such a payback might be sufficient to justify private sector investment.

Waste heat recovery is very case-specific. Each potential application must be evaluated on its own merits to determine if the value of the recovered energy is worth the investment. Rising energy prices certainly focus more attention on the recovery of waste heat, but this does not guarantee that the economics will justify the maximum recovery which technology would permit. The need to consider the economic as well as the technological aspects of waste heat recovery underscores the importance of defining energy conservation as economic efficiency—a point made in the introduction to this book.

PROCESS CHANGES

Process changes by nature occur gradually. They generally are restricted to new plants and equipment because the cost of replacing existing equipment is so much higher than the incremental cost of incorporating the new design into a plant already scheduled to be brought on-line. Process changes therefore normally only occur through the construction of increased capacity or the normal replacement of existing equipment whose service life has ended.

A typical process change would include the use of continuous casting in both steel and aluminum. In both industries the hot metal presently is cooled at least once before it is transformed into a final product.

Most of the steel produced in this country is cast into ingots, allowed to cool, and then reheated for rolling into a product. Continuous casting eliminates the ingot stage and has the potential for significant fuel savings. In addition, it can increase the yield of the steelmaking process by about 10 percent, leading to further product cost reductions.

A study by Thermo Electron Corporation estimated that the use of continuous casting in the steel industry could reduce the energy input per ton of raw steel by approximately 1.4 MMBtu's. This is about 5.5 percent of all fuels consumed in steelmaking. However, the following factors limit the degree to which continuous casting can be expected to penetrate the market.

1. Some grades of steel are not amenable to continuous casting.
2. Metal chemistry, temperature, and operating practices must be closely controlled.
3. Product uses are less flexible.
4. Existing plants have ingot casting facilities which do not need to be replaced.

The use of continuous casting in the steel industry probably will rise over the next decade. In 1973, roughly 5.5 percent of the total national output was produced using these processes, as contrasted with 17 and 21 percent in Germany and Japan, respectively. The growth of continuous casting in the United States will depend largely on the amount of new capacity to be constructed. Given the present state of the steel industry, any projection therefore must be suspect.

The importance of process changes should not be underestimated. Between 1947 and 1974, energy consumption per dollar of value added declined from approximately 116,000 Btu's to 80,000 Btu's, despite declining real prices for fuels. Some of this decline can be attributed to substitution of premium fuels, mainly natural gas for coal. Other factors include a shift in product mix, in which primary goods industries accounted for a smaller share of industrial output. But the primary factor behind this decline appears to be a more efficient use of raw materials, derived indirectly from improvements in existing processes and process changes. For example, the average amount of energy consumed by the aluminum industry in 1947 was estimated to be twelve kilowatt hours (kWh) per pound, and by 1975 had declined to roughly eight kWh per pound. Although the cost of electricity was a significant consideration, equally important was the ability to expand plant capacity and other cost saving factors included in improved productivity.

These issues are important because changes in process design seldom, if ever, will be determined solely on the basis of energy conservation. Only in the case of cement and aluminum does the cost of the energy inputs exceed 10 percent of the final selling price, and for most industries, 2 to 3 percent would be a more common figure. Industries are not going to experiment with energy saving innovations that promise a 10 to 15 percent reduction in energy use (equivalent to a cost savings in the range of 0.1 to 0.4 percent) if the system promises to be any less reliable or more uncertain than the existing design. Energy conservation, or improved energy productivity through process changes, will be limited to the extent that the new processes are compatible with the overall integrity of the production process.

PRODUCT CHANGES

Although product changes and product substitution can result in substantial energy savings, the concepts involved are too complex to be adequately developed in this paper. Product changes involve the redesign of products which consume energy when placed in service, e.g., motors, compressors, etc. The redesign issue is whether these products can be built to specification which reduces their energy consumption characteristics while simultaneously meeting the same level of reliability and productivity when placed in service. Likewise, product substitution involves the use of materials which require less energy to produce but at the same time can result in the same level of consumer satisfaction. Both of these concepts are methods by which scarce resources such as our finite energy supplies can be more fully utilized but which deserve a much more intensive analysis than can be provided in this chapter. Robert Hemphill, in Chapter 4, discusses some of the possibilities for product change in the transportation sector.

AGRICULTURAL ENERGY CONSUMPTION

Much of the above discussion has focused on the manufacturing sector. Although the manufacturing sector accounts for over 80 percent of the industrial sector's energy requirements, the agricultural sector also is important to the economy and in terms of energy consumption.

Direct primary fuel inputs into the agricultural sector account for roughly 2 percent of the nation's energy requirements or roughly 1.5 quadrillion Btu's. These statistics exclude fertilizers such as ammonia.

If fertilizers were included, the figure would be closer to 3 percent or 2.2 quadrillion Btu's.

The principal uses of energy are farm vehicles (50 percent); machine drive, including irrigation, grain handling, and feed processing (25 percent); and crop drying (8 percent). Thus energy consumption in the agricultural sector is dominated by internal combustion engines and electric motors. To the extent that the energy efficiency of these units can be improved, the overall energy productivity of American agriculture can continue to advance. The issues here, however, far surpass just the agricultural sector because diesel and gasoline-powered engines and electric motors are used by every sector of the economy.

The agricultural sector's reputation for high energy consumption stems from its intensive use of heavy equipment for planting, cultivating, and harvesting and its dependence on chemical fertilizers to improve crop yields. Whether these characteristics of agriculture will change because of higher energy prices or scarce fuels will depend on how other production costs change relative to energy costs and availability. The U.S. agricultural sector likely will continue to be more dependent on energy than other nations of the world, and therefore security of supply must continue to be given a high priority.

Potential Energy Savings

The above discussion should serve to illustrate that the concept of energy efficiency is not amenable to simple generalizations. It is extremely complex, involving hundreds of thousands of decision-makers and nearly as many potential opportunities. Despite this complexity, several exhaustive studies have been performed that can provide a rough approximation of the magnitudes of the potential savings. These studies are best discussed in terms of near-term (prior to 1980) and mid-term to long-term (through 2000) opportunities.

NEAR-TERM OPPORTUNITIES

Most near-term opportunities concern the first two categories, i.e., housekeeping and waste heat recovery. Studies performed by the Federal Energy Administration and the Department of Commerce show that the opportunities can be quite significant.

In 1974, the Federal Energy Administration, in cooperation with the Department of Commerce, launched a voluntary program to conserve energy within the industrial sector, appropriately known as the Voluntary Industrial Energy Conservation Program. Initially, the program was restricted to the six most energy intensive industries, but later it was expanded to ten industries.

The basic objective of the program was to obtain from each of these industries voluntary commitments to reduce their energy requirement per unit of output by a fixed percentage by 1980. The final targets and the progress which these firms had made by 1977 are shown in Table 3. The progress of these industries is encouraging and provides some assurance that ultimately they should be able to achieve or exceed their energy efficiency improvement targets. If these industries achieve their goals by 1980, the estimated reduction in energy requirements is equivalent to reducing crude oil imports by 1.2 million barrels per day (MMBD). Table 4 shows the estimated saving for each industry. Since these industries only represent 60 percent of total industrial consumption, savings by industries not represented by the program should result in an even larger reduction. If the excluded industries are only half as successful, the savings will be in the range of 1.6 MMBD, whereas if they do as well, the total impact of near-term opportunities could be in the range of 2 MMBD. The latter number would be equivalent to the 1980 production target for Alaskan crude oil.

MID-TERM TO LONG-TERM OPPORTUNITIES

Knowledge of energy efficiency improvements in the industrial sector does not extend much beyond the 1980 time frame. These savings will be accomplished through better operating and maintenance practices (which

TABLE 3. PROGRESS TOWARD ACHIEVING ENERGY CONSERVATION GOALS

	Standard Industrial Classification (SIC)	*Percent Improvement in Energy Efficiency*	
		Realized 1976	*Target 1980*
Food	20	11	12
Textiles	22	12	22
Paper	26	9	20
Chemicals	28	10	14
Petroleum	29	10	12
Stone, clay, glass	32	8	16
Primary metals	33	4	9
Fabricated metals	34	8	24
Machinery	35	19	15
Transportation	37	13	16
Composite average		8	13

Source: Department of Energy, *Industrial Energy Efficiency Program,* Annual Report, unpublished, March 31, 1978.

TABLE 4. POTENTIAL SAVINGS FOR 1980 [a]

SIC		1972 Energy Use (Base Year) 10^{12} Btu/yr.	1980 Energy Use with Base Year Efficiency 10^{12} Btu/yr.	1980 Energy Use Assuming Attainment of Net Targets 10^{12} Btu/yr.	Projected 1980 Savings through Attainment of Net Targets	
					10^{12} Btu/yr.	1-BFOE per day [b]
20	Food and kindred products	1,047	1,195	1,052	143	62,200
22	Textile mill products	474	567	440	127	55,200
26	Paper and allied products	1,388	1,526	1,210	316	137,400
28	Chemical and allied products	3,087	4,800	4,128	672	292,200
29	Petroleum and coal products	2,993	4,007	3,527	480	208,700
32	Stone, clay, and glass products	1,462	1,753	1,478	275	119,600
33	Primary metal industries	4,246	5,167	4,690	477	207,400
34	Fabricated metal products	442	587	445	142	61,800
35	Machinery except electrical	437	707	601	106	46,100
37	Transportation equipment	414	690	580	110	47,800
		15,990	20,999	18,151	2,848	1,238,400

[a] In terms of total energy use.
[b] Barrels fuel oil equivalent (BFOE) per day; conversion factor is 6.3×10^6 Btu per BFOE.
Source: Department of Energy, *Industrial Energy Efficiency Program*, Annual Report, unpublished, March 31, 1978.

are labor intensive) or through the use of off-the-shelf technologies, which in the past have not offered a high enough return on invested capital. Beyond 1980, many of these self-evident opportunities will be fully exploited and further improvements will be dependent on the development and commercialization of new technologies, process changes, and/or product substitution.

A second issue of importance in the mid-term to long-term time frame is fuel substitution. Fuel substitution generally is directed at substituting relatively abundant fuels—coal, wood, or electricity (generated by coal or nuclear) for scarce fuels—oil and natural gas. This was a major objective of the National Energy Plan and continues to be one of the largest potential means by which the nation can reduce its dependence on imported oil and increase the supply availability of natural gas to small industrial, residential, and commercial markets.

It should be noted that the goals of energy efficiency and fuel substitution are not necessarily complementary. Conflicts arise because, depending on the type of fuel which is being substituted, the process could require more input energy per unit of output. This would happen if a coal-fired boiler meeting stringent environmental regulations were substituted for a natural gas-fired boiler, and the difference in efficiency could be as high as 5 to 10 percent. This should not, however, be interpreted to mean that fuel conversion to less efficient fuels is not desirable. It does, however, point up the problem of focusing solely on Btu's without considering the quality of the Btu's consumed. The second area of potential conflict is the competition for scarce capital resources. Few industries have sufficient capital to pursue every opportunity which promises an adequate rate-of-return, and, in the areas of energy efficiency and fuel substitution where the investments are of a discretionary nature, this conflict is very evident. To a large extent, however, both of these objectives can be pursued with substantial benefits to be derived by both U.S. corporations and the public.

Two studies undertaken by the Energy Research and Development Administration and, most recently, the Department of Energy have focused on industrial fuel requirements in 1985 and 2000. These studies are particularly relevant because they reflect the current National Energy Plan.

1. Energy Research and Development Administration, *The Market Oriented Program Planning Study, Industrial Sector Working Group,* March 1978, and
2. Department of Energy, Office of Energy Technology, *The Industrial Sector Technology Use Model,* June 1978.

Both of these studies focused on improvements in energy efficiency accomplished via new or existing technologies and fuel substitution decisions.

The Market Oriented Program Planning Study (MOPPS) was performed by an intraagency task force representing each of the Energy Research and Development Administration program offices. Baseline price and fuel availability assumptions were supplied by a central advisory committee chaired by the Assistant Secretary for Policy and Analysis. The Industrial Sector Technology Use Model (ISTUM) analysis was an extension of the work performed by the MOPPS task force and was funded by the Division of Energy Technology.

Table 5 provides a comparison between the results of the MOPPS study and the recent ISTUM analysis for advanced conservation technologies currently being developed by the Department of Energy. Although only a subset (40 percent) of the projects which are under development was included in the ISTUM analysis, both studies projected a significant and expanding role for advanced conservation technologies.

It should be noted that in very few instances did the individual technologies included in either analysis promise large savings. It was only through the combination of many diverse projects focusing on specific opportunities that large energy savings occurred. However, to the extent that these projects are successful, the opportunities for significant reduc-

TABLE 5. MARKET PENETRATION ESTIMATES FOR ADVANCED ENERGY CONSERVING TECHNOLOGIES

	MOPPS		*ISTUM*	
Project Category	*1985*	*2000*	*1985*	*2000*
Energy saving technologies				
• Process changes	.636	2.046	.175	.694
• Waste heat	1.592	3.369	.240	.682
Subtotal	2.228	5.415	.416	1.376
Energy using technologies				
• Industrial waste fuels	.959	2.632	.911	2.034
• Advanced cogeneration	.521	2.402	.121	1.693
Subtotal	1.480	5.034	1.032	3.727
Total of all conservation technologies	3.708	10.449	1.448	5.103

Included are 40 percent of the projects being funded by the Division of Industrial Energy Conservation which could be fully documented within the timing and level of effort provided.

tions in the industrial sector's energy requirements are apparent. In addition, these technologies were generally very cost-effective, promising in many cases aftertax rates-of-return far in excess of 15 percent.

Fuel substitution is an equally important question. The most abundant fuel source which can play a major role in satisfying the energy needs of the industrial sector is coal. Whereas the cost of coal on a per million Btu basis is projected to be approximately half that of oil or natural gas, the capital cost of converting coal into useful work energy can be two to four times as expensive as either alternative. Table 6 provides rough estimates of the total cost of generating steam using a coal, oil, or gas boiler. Even though the cost of oil and gas in this example is over 100 percent more than coal, the capital and operating costs of converting coal into steam are more than sufficient to offset the initial fuel price advantage.

The implication of these issues for the future is that unless additional incentives are created for its use, coal will not displace oil and gas as the primary source of energy in the mid term (1985). Figure 2 illustrates how conventional fuels are expected to compete through 2000. This figure only applies to fuels used for heat or power. In the 1974 to 1985 time frame, the dominant fuel used to satisfy new capacity and replacement capacity demands is projected to be oil. Oil captures 4.5 quads out of a total market of 10.5 quads, or roughly 43 percent. Coal increases its market share from 7.4 percent in 1974 to 14.7 percent in 1985 but has less than half the impact of oil in terms of satisfying new demands.

TABLE 6. COMPARISON OF COST TO GENERATE STEAM IN INDUSTRIAL BOILERS (UNIT SIZE: 80,000 LBS. STEAM/HOUR)

	Coal	*Oil*	*Natural Gas*
Input Assumptions			
• Capital cost (10^6\$)	5.0–6.0	1.7	1.6
• Operating and maintenance (\$/1000 lbs.)	134.1	10.6	10.6
• Fuel (\$/MMBtu)	\$1.40	\$3.07	\$3.29
Annualized Cost (10^3\$)[a]			
• Capital	800–960	272	256
• Operating and maintenance	470	37	37
• Fuel	577	1266	1356
Total	1847–2007	1575	1649
\$/1000 lbs. steam	\$5.27–\$5.72	\$4.50	\$4.70

[a] Assuming a 50 percent annual load factor.

Fig. 2. Industrial Requirement for Purchased Fuel and Electricity [a] *(1974–2000)*

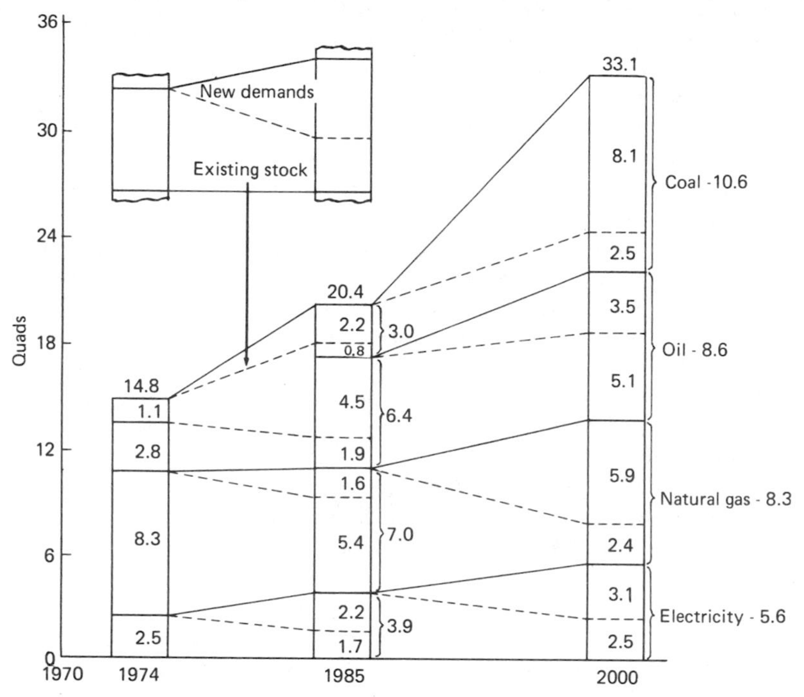

[a] Excludes all oil, gas and coal feedstocks, asphalt, lubes, and waxes, gas and oil consumed in mineral extraction, industrial waste used as a fuel source, and oil used to power agricultural and construction equipment.

The situation changes significantly after 1985. At this point the model assumed that price controls on oil and natural gas would be eliminated and that natural gas supply constraints would be lessened through a combination of synthetics, imported liquefied natural gas, and domestic suppliers' responses to higher wellhead prices. The combination of higher prices and enhanced natural gas supplies (or synthetic fuels) correspondingly increases the competitiveness of coal and reduces the burden on oil to satisfy traditional natural gas markets. Coal, between 1985 and 2000, is projected to satisfy 39.3 percent of the new demands and account for approximately 32 percent of the total demand for heat and power. While the consumption of oil continues to rise (from 6.4 quads in 1985 to 8.6 quads in 2000), consumption decreases as a percent of total demand from 31.4 percent in 1985 to 26 percent in 2000. After approximately one and a half decades of declining use (1970–1985), consumption of natural gas is expected to increase to its 1974 level. Natural gas also captures a significantly larger share of the new demand, increasing from 15 percent in 1985 to 28.6 percent in 2000. In total, both oil and natural gas decline significantly as a share of total demand, decreasing from 75 to 51 percent of total fuels used for heat and power.

Summary and Conclusions

Figure 3 summarizes how the industrial sector can be expected to react to higher priced and less secure fuel supplies throughout the remainder of this century.

During the immediate future, only energy conservation offers any significant promise of reducing the industrial sector's reliance on scarce fuels. Depending on how industries not involved in the Voluntary Industrial Energy Conservation Program react to higher prices, the net savings resulting from the application of existing technology would be in the range of three to four quadrillion Btu's by 1980. Statistics reported by firms participating in the program show a reduction from 13.2 quads in 1972 to 12.6 quads in 1976. This reduction reflects the impact of initial efforts to conserve energy as well as slow growth in industrial output over this period. These statistics, however, also reflect a significant trend away from natural gas to oil. During this period, oil, as a percent of total requirements, rose from 22 to 25.7 percent, while natural gas decreased from 36.8 to 31.9 percent.

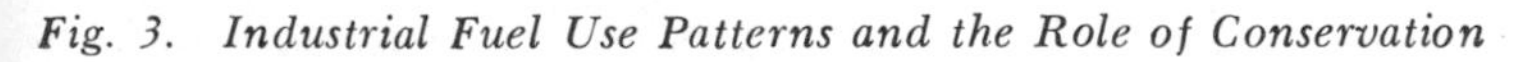

Fig. 3. Industrial Fuel Use Patterns and the Role of Conservation

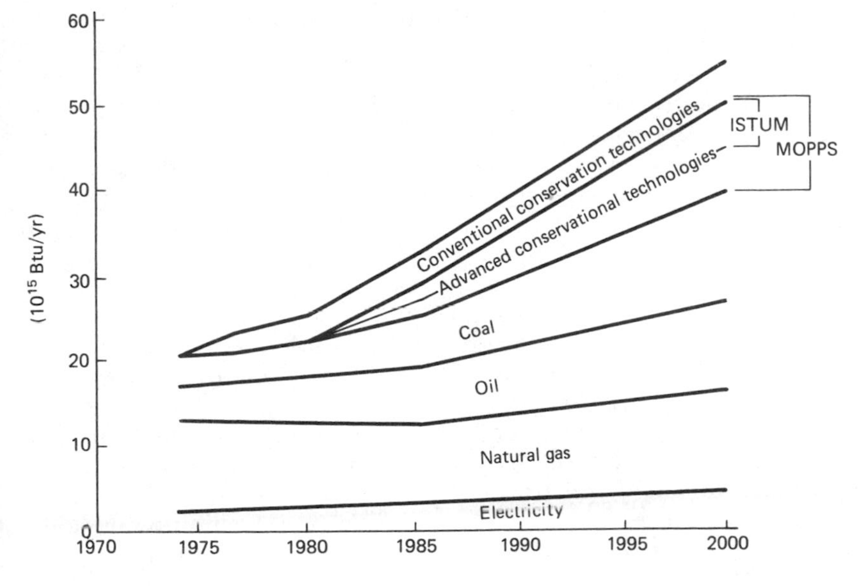

Beyond 1980, opportunities to improve energy production are constrained both by economics and the availability of technological alternatives. A large percentage of the existing options to conserve energy will be exploited by 1980. Without advancements in the design of new processes (e.g., more efficient waste heat recovery techniques and use of cogeneration), it is likely that growth rates in industrial energy consumption could revert to pre-1970 patterns. This will occur during a period in which coal still may cost more to use than oil, aggravating the shift to oil which is already occurring. The impact of advanced conservation technologies is projected to be anywhere in the range of 1.5 to 3 quads by 1985 and 5 to 10 quads by 2000. To the extent that these savings do not occur, the burden of maintaining industrial production will fall heavily on existing fossil fuel supplies. In the 1985 time frame, most of this burden would be absorbed by oil.

Pindyck in Chapter 2 indicates some of the difficulties of estimating price elasticities, and as is true in the consumer sector, we do not know how long it will take the industrial sector to fully react to higher prices. The long run may be ten or fifteen years, or equally likely, it may be thirty to fifty years. Capital stock turnover is highly dependent on many factors other than energy prices and, therefore, cannot be predicted with any degree of certainty. Even if we were to replace our existing capital stock, there is no certainty that we could or even would want to achieve the same relationship between factors of production as reflected by industrial processes in other industrialized countries.

Policy measures to ensure that energy productivity will continue to improve therefore must be directed at providing the opportunity and incentive to improve. These measures generally can be discussed in the context of those which encourage technological development and those which influence economic decisions. It is important to recognize that the latter is only relevant if the first occurs.

Since energy conservation was not an important consideration prior to the 1974 rise in energy prices, very little research in this area was conducted during the 1950s and 1960s. Most of the opportunities being exploited by the industrial sector have been known for decades but only recently proved to be economic. In economic terms, unless new technologies are developed, the marginal cost function for energy conservation promises to be very steep in the 1980s which, unless energy prices double again, will result in slowing the rate of improved energy productivity substantially.

The first policy to be pursued, therefore, is to ensure that adequate research is conducted to develop, demonstrate, and commercialize concepts that ultimately will permit a continuous improvement in energy productivity. This process can be one of joint participation by the public and private sectors. Since the time frame for research by the public sector can be substantially longer than in the private sector, the respective roles of the sectors are different. Theoretically, the fragmented nature of many industries lacking the resources to fund research projects individually, or constrained by antitrust considerations from pursuing joint programs, may force the public sector to play an extremely significant role.

However, while necessary, it is not enough to ensure that new conservation technologies are developed. For years the United States has valued its energy supplies below the true replacement cost. Today our only alternative source of supply is increased oil and natural gas imports. Holding prices below world prices leads to increased consumption and lower domestic supplies. It also tends to discourage fuel substitution which, in this case, favors increased oil over coal consumption. At a minimum, it must be made clear that decisions regarding energy consumption within the industrial sector should be made based on the replacement cost of energy to the nation. In the near term, this could mean pricing energy supplies at the cost of imported oil or natural gas. We pay roughly fourteen dollars a barrel for imported crude oil or $2.30 per MMBtu; the cost of imported liquefied natural gas is expected to be closer to three to four dollars per MMBtu. Over the long term, the replacement cost could be defined as replacing oil and gas supplies with domestically-derived substitutes. These substitutes might include synthetic fuels derived from coal, estimated to cost from four to six dollars per MMBtu, or solar energy, which is expected to carry an even higher price tag. Regardless of the replacement cost actually used, the nation needs to realize that its oil, gas, and coal supplies are a finite resource which, once exploited, can be replaced only through very expensive alternatives. These alternatives should be valued not only in terms of dollars and cents, but also in the context of U.S. vulnerability to future embargoes, restrained foreign policy options, and the environmental degradation that will occur as a result of being forced to make full use of our coal reserves.

The energy needs of the industrial sector must be met. Alternative technologies and sources of supply must be developed and exploited as rapidly as possible. Improved energy productivity and increased use of coal are necessities both in the context of maintaining the economic viability of American industry and ensuring adequate supplies of premium fuels for the other sectors of the economy which have fewer choices.

Douglas Bauer and
Alan S. Hirshberg

7

Improving the Efficiency of Electricity Generation and Usage

Introduction

Previous chapters have discussed the relationship between energy consumption, energy prices, and the U.S. GNP. Pindyck has argued that energy demand is more elastic than some current estimates (at least in the long run), and Hogan has shown that if these higher estimates for elasticity (0.7 versus 0.2) are valid, then the long-term impact of higher energy prices on the GNP will not be as large as some fear. Substitutions of capital (and perhaps labor) will be made for higher priced energy, and these substitutions will prevent large-scale economic damage.

Even if substitutions are likely in the long run, two problems still remain. First, what policies should be adopted during the next ten years; i.e., what set of policies is necessary to accomplish optimal substitutions while keeping short-term GNP effects to a minimum? Second, how shall the substitutions be implemented; that is, what substitution options exist and how can they be managed to provide the information which will

---

DOUGLAS BAUER *is assistant administrator in charge of utility systems for the Economic Regulatory Administration. Prior to that he was director of the Division of Nuclear Research in ERDA. In 1972–73 he was special assistant to the secretary of transportation.*

ALAN S. HIRSHBERG *is research director with Booz, Allen and Hamilton, Inc., Bethesda, Maryland. Previously, he was a project manager at California Institute of Technology, Jet Propulsion Laboratory, and president of Environmental Future, Inc. in California.*

produce the necessary substitutions as quickly as possible? The electric utilities provide one major opportunity for energy reductions by substituting more efficient "conversion" equipment and "utilization" patterns into the energy use picture. The purpose of this chapter is to discuss, in general terms, the range of efficiency increasing options available to utilities and to identify means by which the efficiency of electricity usage can be improved in a cost-effective manner.

Efficiency, as used in this chapter, is a generic concept, with meaning much broader than its usual engineering definition. In this context, efficiency encompasses the engineering concept of generation efficiency (e.g., improved heat rate of generation equipment), as well as utilization efficiency which can be accomplished by shifting a portion of electricity use from times of peak demand to off-peak periods, thus reducing the total cost of electricity service and, thereby, increasing the utilization of base-load capacity. Efficiency also implies the use of more efficient end use devices, but since that area is treated elsewhere in this volume, it will not be discussed in this chapter.

This chapter is organized into four major sections. To provide a context for understanding the gains that could accrue from improvements in efficiency, the first section is devoted to an abbreviated discussion of some of the key characteristics of the electric utility industry. In addition to a discussion of the structure and regional variation of the industry, the first section describes the aggregate national trends of electricity usage and three of the major problems that industry faces—uncertainty in the growth of electricity demand, difficulty in selection of new capacity, and the industry's financial problems. The second section discusses the types of efficiency improvements which could help reduce some of the problems which the industry faces. The third section describes the options which are available for improving both conversion and system utilization efficiency. The fourth section discusses the institutional barriers which may delay or prevent these improved efficiency options from being adopted. The last section will summarize the major conclusions of the chapter.

Characteristics of the Electric Utility Industry

The electric utility industry is both complex and diverse. Financial conditions, load characteristics, fuel costs, and equipment mix vary widely from utility to utility. Generalizations about the industry are largely invalid, and the efficacy of options for improving efficiency must be evaluated at the individual system level.

The electric utility industry is comprised of more than 3,500 companies. Most electric utilities operate as local monopolies for the purpose of generating and/or distributing electricity. Such monopolies eliminate

Fig. 1. The Electric Power Industry—1975

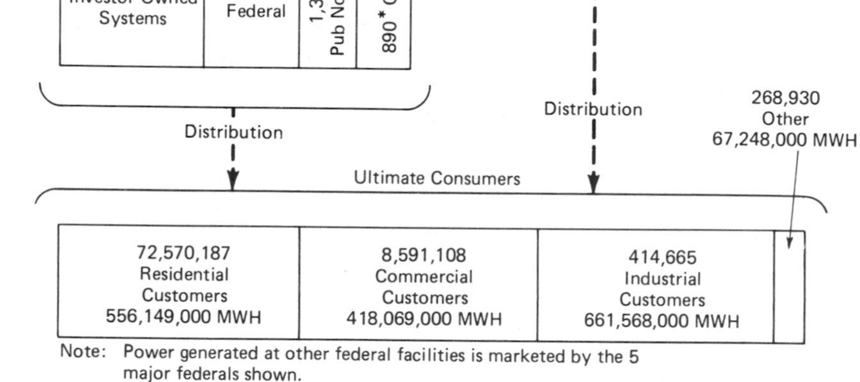

Note: Power generated at other federal facilities is marketed by the 5 major federals shown.

*Estimated

Source: Booz, Allen and Hamilton, Inc.'s report to Office of Technology Assessment, *The Utility Role in Conservation,* December 1977.

expensive duplication of generation, transmission, and distribution equipment. Figure 1 depicts the structure of the industry as it existed in 1975 and shows how the generation, transmission, and distribution segments are linked among privately and publicly owned systems.

Investor, or privately owned, utilities provide the bulk of electricity to residential customers. In 1975, about 250 investor owned electric utilities accounted for about 80 percent of the industry's total generating capacity. Additionally, 150 smaller investor owned utilities were engaged solely in the distribution of electricity.

Cooperatively owned electric utilities supply power to many rural areas of the country. Most of the approximately 1,000 rural electric cooperatives are small, and over 90 percent are engaged solely in the distribution of electricity. These systems purchase substantial amounts of their energy for distribution and resale from federal or other private systems.

In 1975, there were 2,245 municipally owned electric utilities in the U.S. In 1975, their sales accounted for about 9 percent of total industry sales. Many of these utilities purchase power for resale from municipal wholesalers and federal power authorities. The municipal wholesalers are themselves included in this category (e.g., Power Authority of the State of New York). The wholesalers supply electricity to municipal systems, rural electric cooperatives, and private utilities who in turn resell the electricity to retail customers.

The federally owned sector of the utility industry supplies power for resale and distribution and accounts for about 10 percent of the nation's installed capacity. Additionally, these systems provide approximately 5 percent of the total sales to ultimate consumers (including industrial and commercial businesses). With the exception of the Tennessee Valley Authority, these utilities (including the Bonneville, Alaska, Southwestern, and Southeastern Power Administrations) are part of the Department of Energy. Each of these systems is required to have its rate schedules approved by the Economic Regulatory Administration within the Department of Energy.

Electric utility characteristics tend to be regional and system-specific. While the national trends are important to understand for energy policy planning, understanding regional differences is crucial to the implementation of national plans. Failure to adequately account for regional differences could seriously impede national efforts to achieve efficiency gains.

As indicated in Figure 2, each region of the country has unique usage and generation patterns. In 1976, coal-fired and oil-fired capacity accounted for over 57 percent of all generating capacity. However, abundant hydropower (63 percent) in the western states, particularly the Northwest, made electricity in this region cheaper than in other parts of

Fig. 2. 1976 Generation Capability by Reliability Council and the Location of Each Regional Reliability Council

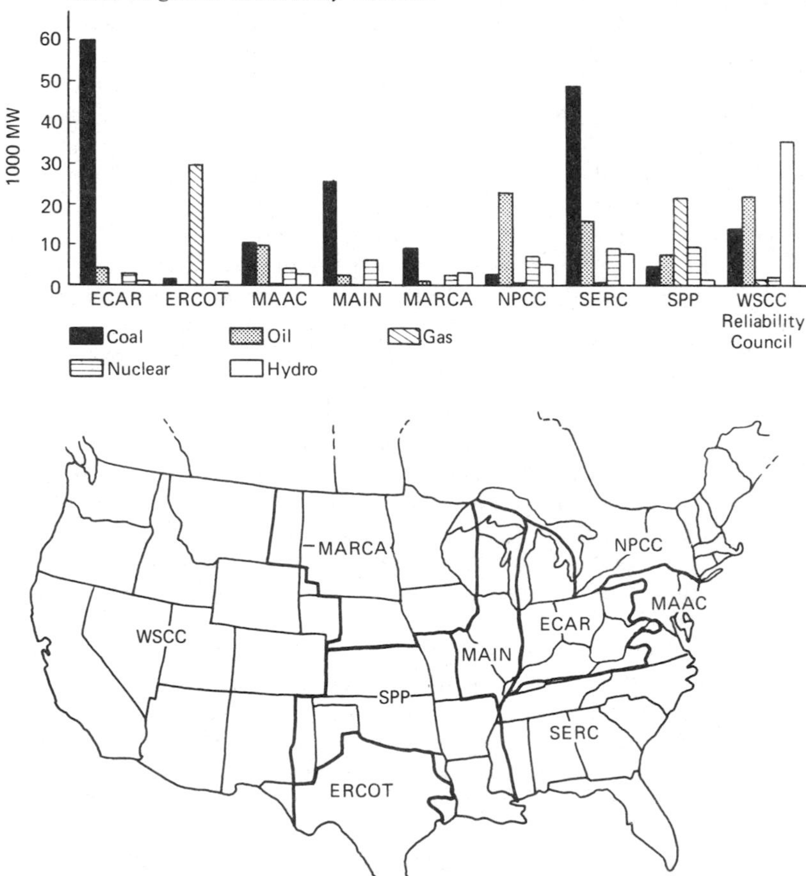

Identification of Reliability Councils:

ECAR—East Central Reliability Coordination Agreement
ERCOT—Electric Reliability Council of Texas
MAAC—Mid-Atlantic Area Council
MAIN—Mid-Atlantic Interpool Network
MARCA—Mid-Continent Area Reliability Coordination Agreement
NPCC—Northeast Power Coordination Council
SERC—Southeastern Electric Reliability Council
SPP—Southwest Power Pool
WSCC—Western Systems Coordinating Council

Source: 7th Annual Review, National Electric Reliability Council, July 1977, p. 5 (Note: Data from the *8th Annual Review* indicate relatively small changes from 1976 to 1977.)

the nation. The large percentage of oil-fired and gas-fired capacity in the regions which contain the larger gas and oil producing states of Texas and Louisiana kept prices relatively low in these regions too. However, over the next decade, capacity additions of these types will be limited and will alter the fuel mix in all regions.

The regional growth rates in electricity differ significantly from national averages. While the heavily industrialized eastern regions currently produce a large fraction of U.S. electricity, the southern and western regions of the country are growing at faster rates and will alter this distribution. The Northeast, which is extremely dependent upon expensive imported oil for its electricity generation (57 percent of New England's electricity is generated by oil, and approximately 90 percent of that oil is imported), is currently experiencing a growth rate of about 4 percent. In contrast, demand in Texas (an oil and gas producing state and part of the "sun belt") is growing at approximately 8 percent annually. Much of this electricity has been generated with natural gas, but shifts are underway to use Texas lignite, Montana coal, and nuclear power.

Forecast increases in annual kilowatt-hour requirements also differ among various regions. The Northeast is forecast to experience a relatively low growth rate in kilowatt-hour requirements through 1986, due in part to its slow economic growth rate compared to the rest of the nation. From 1976 to 1987, the growth of kilowatt-hour requirements is predicted to be approximately 4.4 percent in this region, while growth in other parts of the country should average around 5.3 percent. However, states in the Southwest are forecast to experience growth in kilowatt-hour requirements of 6.8 percent, possibly because population and industrial use are growing rapidly.

Electric utilities not only supply one of our most important and versatile forms of energy, but consume a large portion of our primary fuels (oil, natural gas, coal, uranium). In 1978, electricity generation consumed about 30 percent of all primary fuels used in the U.S. Fuel for electricity generation required about 70 percent of the coal produced in the U.S., petroleum equivalent to 20 percent of domestic production, and about 13 percent of domestic natural gas. Although (as shown in Figure 3) only about 13 percent of 1978 electricity was produced by fission of uranium in nuclear reactors, this share is projected to increase during the next decade so that by 1987 approximately 27 percent of electricity will be generated from nuclear plants. While the contribution from nuclear energy will grow dramatically, coal will continue to be the primary fuel source of electrical generation for the remainder of this century.

During the past decade, the electricity industry has been one of our fastest growing users of primary fuels. Sales of electricity grew from 830 billion kilowatt-hours (kWh) in 1963 to 1,700 billion kWh in 1973—or at about 7.5 percent annually. In order to keep pace with this rapid

Fig. 3. Electric Energy Generation by Principal Sources (Contiguous U.S.)

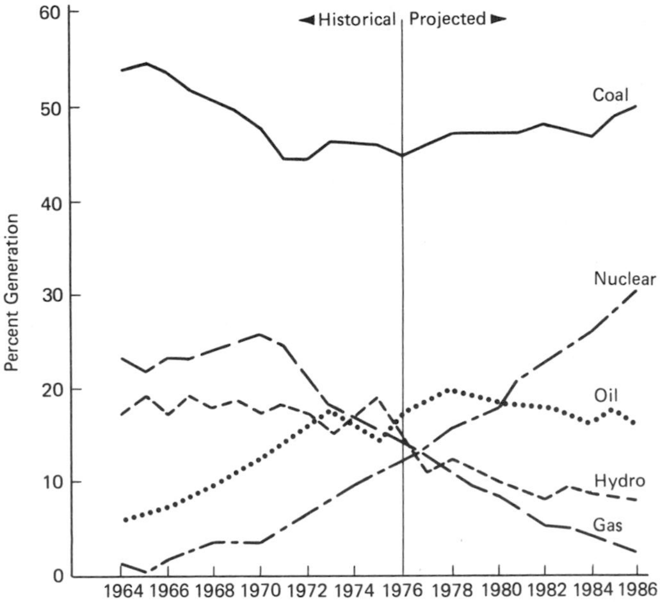

Source: National Electric Reliability Council, *7th Annual Review,* July 1977.

increase in demand, electric utilities have had to double generating capacity about every ten years. Installed capacity has risen from just over 200 gigawatts in 1966 to about 500 gigawatts in 1976. This growth has required large amounts of capital; annual capital expenditures by electric utilities have grown in recent years to an average of $13 billion.

Until the 1970s, the steady growth in electricity sales, the relative ease in obtaining financing for new generating plants, and the steady or declining cost of new sources of electricity made the electric utility industry one of the most stable in the U.S. With the advent of environmental laws requiring expensive pollution controls, federal policies to reduce oil use, public reaction against nuclear generating facilities, and the rapid increase in cost of new capacity, the decision environment for electric utilities has become very unstable. Utilities are faced with escalating uncertainty regarding three of their key decision variables.

First, the *growth in demand* for electricity has been erratic since 1973. The lower growth rates in electricity sales may be either a temporary aberration or the early signs of what Pindyck argues will be the long-term elastic response to higher electricity prices. Because it takes at least ten to twelve years to bring major baseload plants on-line, utilities must make commitments now for projected generation needs a decade away. Even small errors in projecting electricity growth rates will telescope into large differences in required generating capacity ten years later, producing economic inefficiency either because utilities will have too much installed capacity or too little, forcing reliance upon expensive peak power generating facilities which typically require scarce fuels.

Second, the choice of *generation alternatives* has become clouded by a host of issues, usually beyond the direct control of the utility. There are problems with expanding the use of each type of new generating capacity.

1. Hydroelectric power, which supplies about 10 percent of the nation's electricity, is limited by the availability of suitable dam sites and growing environmental opposition to dam construction. Costs of hydro are escalating very rapidly.
2. Natural gas-fired and oil-fired plants are being restricted by public policies which limit or prescribe the use of these fuels for electricity generation.
3. Coal, which accounted for about 48 percent of the primary energy source inputs to electrical generation in 1978, has environmental, transportation, handling, and institutional problems associated with its production and use, all of which will tend to increase the cost and threaten the reliability of coal-based systems.
4. Nuclear power suffers from uncertainties regarding waste disposal, escalating costs for uranium, and public interest group opposition to nuclear power plants.

Third, *capital costs of new plants* have risen sharply, an escalation which has been exacerbated by the slippage in plant construction schedules, due in part to delays in siting plants. Since 1973, the cost of energy kilowatts of added capacity has escalated dramatically, as illustrated in Figure 4. The current lead time for nuclear plants is ten to fifteen years, and cost increases and interest during construction are estimated to account for about half of total plant costs. According to the Department of Energy, average nuclear plant capital costs are expected to double in the next seven years (from the $530 per installed kilowatt in 1977), shown in Figure 5. A compounding problem is that initial plant cost projections have been underestimated. In a recent study of this problem, the American Enterprise Institute noted that three nuclear plants completed in the 1974–75 period had overrun their initial estimates by an average of 49 percent, even after accounting for unanticipated inflation. The financial problem is exacerbated on the one hand by regulatory delays, and on the other hand by the escalation in interest rates which utilities must pay for capital. Capital costs for con-

Fig. 4. Nuclear Plant Capital Costs—Estimate and Actual Comparisons

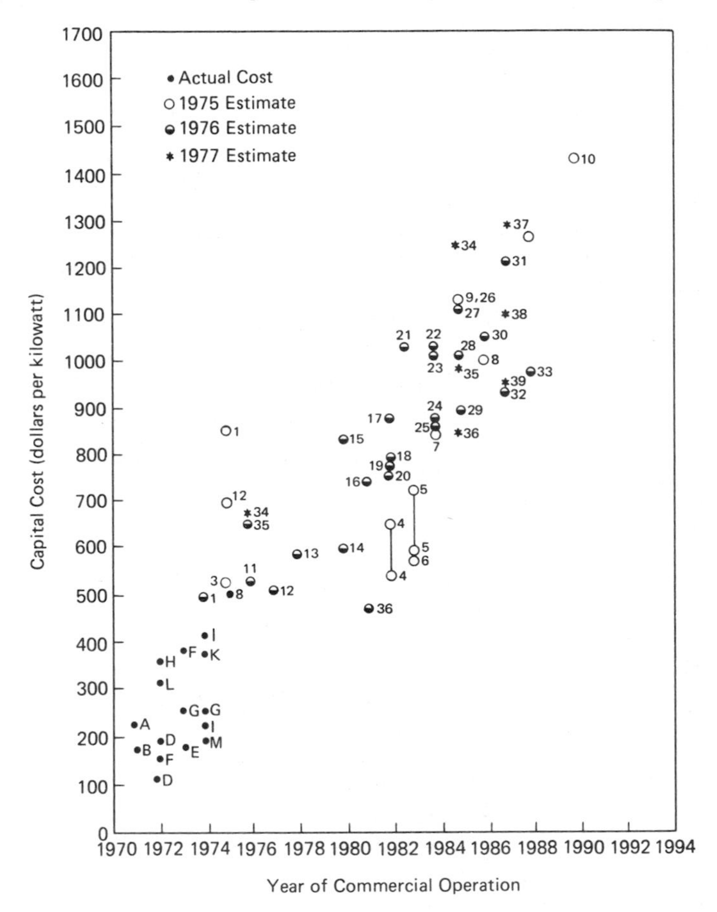

Source: Booz, Allen and Hamilton, Inc.

Letters refer to 13 projects completed prior to 1976.
Numbers refer to estimate for 39 new projects.

Fig. 5. Average Nuclear Plant Cost by Year of Commercial Operation

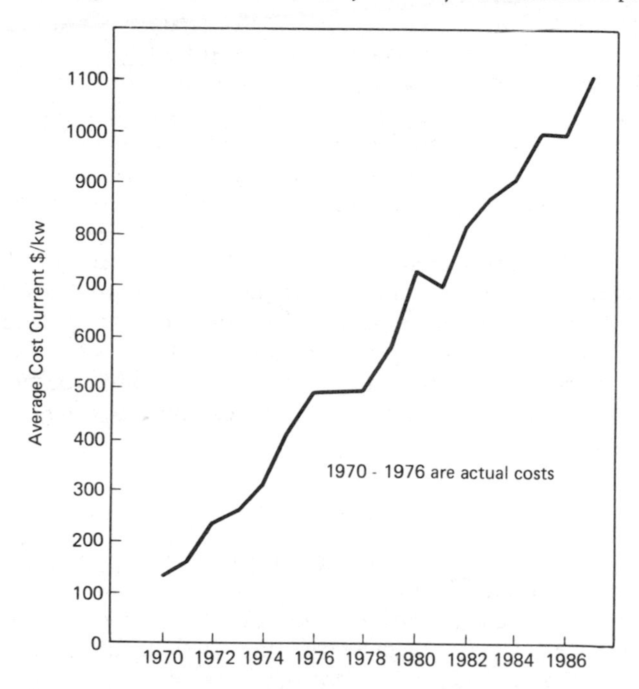

Source: Power Engineering, April 1978, p. 65.

ventional steam generation plants have also risen dramatically since the Arab oil embargo. The cost of a new coal-fired plant has risen from $100 per installed kilowatt in 1967 to $450 per kilowatt in 1978. Over one-third of this cost is caused by clean air compliances, added site preparation, and interest during construction.

One way of reducing the impact of this unstable decision environment is to postpone the need for major capacity additions by improving the efficiency of electricity usage. Improvements in conversion and utilization efficiency could, in effect, "stretch out" the primary resource base and defer the addition of expensive new generating capacity. Even small efficiency improvements can significantly reduce the demand for primary fuels and reduce the near-term capital requirements of the industry. (Capital requirements may ultimately increase as expansion of capacity is delayed, if the rate of inflation in plant costs exceeds the rate of load growth.)

Improving net generating efficiency from 32 to 35 percent would have saved the thermal equivalent of about 900,000 barrels per day of oil in 1977. This savings represents approximately one-third of a 2.5 million barrel per day oil import reduction goal, assuming all of the saving occurs in plants which use imported oil for electricity generation.

Increasing the average utility load factor would utilize existing generating capacity more efficiently and postpone the need for new capacity. If some electricity demand could be shifted from time of peak demand to off-peak demand, utilities could reduce the need for additional capacity. The latest National Electric Reliability Council (NERC) estimate for national annual electricity capacity growth is 5.2 percent per year for summer peak and 5.4 percent per year for winter peak.

Reduction in the energy growth rate from 5.3 to 3 percent would save about 670 billion kWh over the 1977–1987 period (assuming a constant load factor), or about 30 percent of the total 1978 electricity sales.

Before discussing the specific options which could be used to achieve these efficiency gains, it is important to discuss the different meanings of the term "efficiency." The next section describes the meaning of efficiency as used in this chapter.

Types of Efficiency

Theoretically, the electric utility sector has many opportunities for improved utilization of energy inputs. For discussion purposes, two types of efficiency can be defined that are of relevance to electrical energy production and use—"conversion" efficiency and "system utilization" efficiency. Opportunities for efficiency gains are present in each type.

Conversion efficiency refers to the amount of electricity produced per unit of input energy and currently averages about 32 percent across the industry. That is, about one-third of the input energy is converted to electric energy, the other two-thirds is "lost." Of course, dramatic improvements, e.g., beyond 50 to 60 percent, are theoretically impossible because of the laws of thermodynamics; however, several new technologies offer promise of substantial improvements in conversion efficiency from the current industry average of about 32 percent. The essence of these technologies is to utilize the "waste" heat from conversion processes, e.g., expansion through a turbine or process heat or district heating applications.

System utilization efficiency refers to how completely and continuously a utility uses its plant capacity. The efficiency with which a particular utility system's different plant capacity is operated will determine how efficiently the capital investment of a utility system is utilized. The sharp time variation in customer demand patterns had led utilities to the creation of a "generation mix" of plants of differing capital costs, operational characteristics (e.g., start-up time), efficiencies, and fuel dependencies within the same utility. Plants are "dispatched" to meet time-varying loads in three broad categories:

1. *Baseload plants* (typically the most efficient, relying upon coal or nuclear fuels) have the highest relative capital cost but lowest relative operating cost (considering fuel and capital components); base plants meet the base-load component of demand that remains constant throughout the day. For maximum efficiency, these plants will have the highest capacity factor targets.
2. *Intermediate-range plants* which may rely upon coal and oil are more costly to operate than baseload plants. They have slightly lower capital costs, but have relatively higher operating costs. They are, in general, "cycled" or operated for ten to fourteen hours each day.
3. *Peaking plants* which rely on oil or gas are the least efficient of all the plants on a utility system. They typically have the lowest capital costs, but the highest operating costs and, overall, are the most expensive plants to operate on the utility system. They are used to supply electricity during hours of peak demand, typically lasting only a few hours daily. They may also serve as standby or reverse capacity to be used during times of maintenance on other plants because peaking plants can be brought on-line quickly.

Two factors, often confused, are measures of utilization efficiency.

1. *Capacity factor*, defined as gross electricity (in kWh) generated in a year divided by maximum dependable capacity times 8,760 (the number of hours per year), is a measure of the utilization of a particular plant or an entire utility system. Typically, high capacity factors for baseload plants (which have high capital and low operating costs) and low capacity factors for peak load plants (which have low capital and high operating costs) indicate efficient utilization.
2. *Load factor*, which is defined as average load divided by the system peak load times 8,760, measures the efficiency of a system due to the time variation and magnitude of electricity demand. Historically, consumers have had the freedom to choose how much electricity they will use and when they will use it. Traditional rate structures have not encouraged residential customers to use electricity preferentially at times when it is less costly (and more efficient) to supply it, so that peaks and valleys in consumption from season-to-season and during each day have occurred based on consumer preferences and not on generation economics. Peaks usually occur for several hours during the afternoon in summer (for summer peaking utilities), and during the morning and early evening hours during the winter (for winter peaking utilities).

Improvements in both "conversion" and "utilization" efficiencies could play an important role in easing the problems which utilities currently face by increasing the electricity generated from each Btu of primary fuel, by reducing the need for new generating capacity investments, and by improving the productive use of existing capacity. Electricity conversion and utilization efficiency should be key areas of concentration in any national program to use primary fuels more efficiently, to shift reliance from oil and gas to coal, and to reduce the strains of high levels of financing on utilities. The next section describes several options for improving conversion and utilization efficiency.

Efficiency Options

Improvements in electricity usage can be achieved in several ways. Most direct is improved efficiency in generation through technologica improvement. Other methods of improvements concentrate on how efficiently electricity is used. Savitz and Hirst in Chapter 5 argue, for example, that substantial residential energy savings could occur with attendant reduction in household fuel bills if consumers switched to more efficient appliances, such as advanced electric heat pumps and highe efficiency air-conditioning systems. Utilization efficiency improvement in electricity usage would result from a better matching of the demanc pattern with the generating characteristics of the utility. These option have been successfully used in Europe (as discussed by Schipper in Chapter 3) and have been the subject of sixteen experiments since 197 in this country in which 18,000 consumers have participated. Before discussing these utilization efficiency options, it is useful to review possibl improvements in the conversion efficiency of generating plants.

CONVERSION EFFICIENCY OPTIONS

Conversion efficiency improvements have historically been derive from engineering and materials improvements in generating and trans mission equipment. In 1920, the most efficient plants had coal-to-electri city conversion efficiencies of roughly 17 percent. By the middle of th 1960s, super-critical coal-fired plants were being designed with efficiencie as high as 39 percent. Until the late 1960s, improvements in material and technology allowed utilities to achieve higher conversion efficiencie using conventional Rankine steam cycle equipment while capturin economies of scale. At that time, however, it became apparent that ob taining higher efficiency would require exotic, expensive materials and/o risk imposition of unacceptable reliability penalties on the units.

Contemporary, reliable generating technology now appears to hav reached a plateau both in size and efficiency, and large improvements i conversion efficiency must derive from the development of other ad vanced conversion or transmission improvements, e.g., magnetohydro dynamics (MHD) or extremely high voltage (EHV) transmission. By th 1960s, bulk power lines operating at 765 kilovolts permitted large powe exchanges at high efficiency over hundreds of miles, capturing system load diversity advantages; but a pause now exists as the environmental effects are debated and researched.

After the passage of NEPA, environmental considerations began to impose sharply increased design constraints on power plants. Particulate and sulfur control devices to satisfy "best available technology" criteria have reduced both conversion efficiencies and operating reliability, while increasing generation costs.

Electric utilities have at least three options for reducing their primary fuel consumption while generating the same amount of electricity.

1. *Developing new generation technologies,* thereby improving the conversion efficiency of central station production.
2. *Switching to solar and other renewable resources,* thereby conserving scarce primary fuels for uses where substitution is less feasible.
3. *Recovering waste heat from one conversion process for other uses,* thereby increasing the level of amenities provided for the same input of primary fuels.

Developing New Generation Technologies—A large number of conversion cycles that would refine existing technologies or present new generation alternatives have been identified. Ongoing research and development programs sponsored by the Department of Energy, the Electric Power Research Institute, and industry are exploring many of these options. Most prominently considered are:

1. *Coal-based systems*
   Phosphoric acid fuel cells,
   Fluidized bed combustion,
   Combined cycles,
   Molten carbonate fuel cells,
   Magnetohydrodynamics.
2. *Nuclear-based systems*
   LWR improvements to extend burn-up, thereby conserving uranium and separative work units (SWU) which would be lost in conventional approach;
   Gas-cooled reactors operating at higher temperatures and efficiencies and exploiting thorium fuels;
   Liquid metal fast breeder reactors;
   Fusion.

Table 1 presents an overview of the status of and prospects for these systems. Two points deserve mention in connection with this table.

First, the date of first commercial application is the authors' estimate based upon estimates made by researchers and program managers in-

Table 1. Major Advanced Coal-Based and Nuclear-Based Generation Options

Technology	*Conversion Efficiency*	*Projected Date of First Commercial Use*	*Key Hurdles to Commercialization*
Fuel cells (phosphoric acid)	35–40%	1983–1986	Projected cell life and economic goals must be demonstrated. Requires an extremely clean fuel (such as natural gas or naphtha)
Fluidized bed combustion	38–44%	1985–1995	Boiler configuration and materials problems must be overcome. Limestone (sorbent) use has been greater than anticipated
Coal-fired combined cycle	40–45%	1990–1995	Advanced components must be developed, tested, and integrated at a utility-scale demonstration
Fuel cells (molten carbonate)	45–50%	1993–1998	Second-generation cells are specifically aimed at utility markets. Coal-derived fuels may be used, but much R&D will be required before technical and economic targets can be demonstrated
Magneto-hydrodynamics (MHD)	50–60%	Post-2005	MHD is in a very early stage of development. The technology has tremendous long-term potential, but many fundamental engineering problems must be overcome
Liquid metal fast breeder reactor (LMFBR)	35–40% (Breeds more fuel than it consumes)	Post 2000 in U.S. (Demonstrations on-line in other countries)	Non-proliferation issues and other non-technical barriers make domestic use of the LMFBR highly uncertain
Fusion	Unknown	2010–2020	Fusion is still undergoing basic scientific research. Commercialization problems are unknown at present
Light water reactor (LWR) improvements	15% on fuel	1980's	Unsteady government funding and storage/reprocessing policies
Gas-cooled reactors (GCR)	20% on fuel	1990's	Commercial barriers

volved in each technology. It assumes that no insurmountable problems will crop up and that the required breakthroughs will occur in an orderly manner on the basis of the R&D program planning activities. Each technology is assumed to pass through successive pilot and demonstration plant phases. The commercial application date is the date at which the first commercial plant is projected to actually begin producing electricity.

Implicitly, it is asumed that new technologies will turn out to be economically superior to existing options at the time commercial decisions must be implemented. It should also be noted that a significant contribution from each technology will not be felt until many years after this date; light water reactors were first commercially applied at Shippingport, Pennsylvania, in 1957, but did not account for 10 percent of domestic generation until 1975 despite massive federal support of precursors in the nuclear power program. Utilities are less financially able to assume the risks of commercializing a new technology now than they were in 1957. Commercialization of expensive new technologies will probably require consortia of utilities and manufacturers to share risks. The complexities associated with consortia formation may retard commercial deployment.

Second, technologies with the highest conversion efficiencies are typically the furthest from commercialization, so that dramatic improvements in the efficiency of total electric power production will not take place for many years.

Switching to Renewable Sources—A second option which electric utilities might use to reduce their consumption of primary fuels is for utilities to generate power using renewable resources. A number of distinct technologies are under development, as shown in Table 2. The common problem confronting utility application is end use economics, exacerbated by the fact that most renewable resources are not concentrated, but diffuse. Solar thermal, photovoltaics, and wind power rely on the instantaneous availability of the sun or wind, which poses an additional problem for utility application, requiring either added storage or reserve capacity on the utility system.

Research and development efforts will undoubtedly demonstrate that some of the advanced technologies have insufficient technical promise to warrant commercial introduction, and others may not penetrate the commercial market for institutional or economic reasons. Nonetheless, exploration of generating technologies which use renewable resources is warranted because of their high potential in the long run and because of lack of research on these technologies (compared to other new technologies).

Recovering Waste Heat—The third option noted above was to improve fuel utilization by increasing the useful energy which can be

TABLE 2. MAJOR RENEWABLE RESOURCE GENERATION OPTIONS

Technology	*Date of First Commercial Use*	*Key Hurdles to Commercialization*
Solar thermal	1985–1990	Technical problems center around mirror and collector design. Economics will limit the extent of commercial use.
Photovoltaics	Small-scale units currently available	Achievement of projected cost reductions
Wind	Small-scale units currently available	Reliability and economics are the key issues
Biomass	Current	Availability and economics of collecting sufficient refuse or plant matter
Ocean thermal energy conversion (OTEC)	1990–2000	Requires tropical ocean. Component fouling and economic feasibility pose key problem areas. Low efficiency requires enormous heat exchange areas.

extracted from a given quantity of fuel through utilization of waste heat from one conversion process as an input for a second process. Waste heat recovery technologies include cogeneration, district heating, and total energy systems. Each of these options reorganizes the way in which energy production and consumption activities would take place in the U.S. Both cogeneration and district heating integrate electric generation with other energy consumption activities. Cogeneration links the power plant with industrial facilities which make use of steam bled from the turbine for direct thermal uses. Similarly, district heating uses utility raised steam to supply the space heating requirements of a community. The guiding principle of total energy systems is similar to that of cogeneration; organizationally, however, they differ in that in the latter, the industrial facility may produce and consume all of the energy rather than teaming with a utility partner.

UTILIZATION EFFICIENCY OPTIONS

In addition to the possibility of improving efficiency through advanced or improved generation and transmission technologies, there are opportunities for achieving greater efficiency in the way a utility system is used. These solutions employ cost-effective substitutes of material and labor for energy to provide the same level of amenities. In most homes in the country, for example, increased insulation levels will reduce energy used for space heating while maintaining comfort levels. These options have been designated demand side options and fall into three broad categories:

pricing strategies, load management, and end use conservation (appliances or processes).

The first two options will be discussed in the following paragraphs. The third option, end use conservation, is discussed in other chapters of this volume (see, for example, Chapter 6 by Reid and Chiogioji and Chapter 5 by Savitz and Hirst).

Pricing strategies involve restructuring rates to accurately track the time-varying costs of providing electricity service, thereby sending accurate cost signals to the consumer. Since the costs of generating power at the peak are greater than those off-peak, and some price elasticity exists, usage will shift to off-peak hours when more efficient baseload equipment will supply the electricity demanded. In past years, utility rates have been based upon the average cost of providing service from all facilities on the utility system, regardless of the time the service was provided (nontime differentiated accounting costs [NTDAC]). Experiments with other types of rates have been attempted, including the following.

1. Time differentiated accounting cost (TDAC) rates may be determined similarly to NTDAC rates but with the added step of allocating costs to specific time periods.
2. Time differentiated marginal cost (TDMC) rates are based on a fundamentally different concept than either of the accounting cost-rate designs. The cost of providing peak service (pumped storage, gas turbine) is the basis for energy charges during peak periods. TDMC methods use the cost of *new* capacity as the basis for the rate charge, and the resulting costs usually exceed those derived using accounting-cost methods. The application of marginal cost concepts to utility rate structures requires development in each utility setting to solve a number of problems including excess revenue difficulties.
3. Inverted rate structures charge more for each additional discrete unit of electricity consumed above an established level.
4. Lifeline rates pose a minimal charge for some subsistence level of electric usage for low-income users or all residential users.
5. Demand charges, currently applied mainly in the commercial and industrial sectors, involve higher charges per peak power consumption.

Although the magnitude of customer response to the types of rate design is still largely uncertain, experiments conducted in several areas of the country have indicated that time-of-day rates do produce shifts in the usage patterns of all sectors.

Load management options include other than rate methods to shift consumption to off-peak hours, and methods to reduce peak demand. Shifting consumption to off-peak hours can be accomplished either by storing off-peak energy for use on peak or by changing the *time* of consumption away from peak hours, with the incentive in both cases being lower cost of electric energy. In the former case, however, the incentive to the customer must be large enough to offset the cost of buying and installing energy storage systems or control equipment to regulate, for example, the time electricity heats water in a storage system. In the latter, the customer must perceive economic benefits that warrant his changing *when* he uses electricity, e.g., he may defer the use of electrically operated clothes drying to a time when the utility is not experiencing peak demand.

Load management options fall into three broad categories—voluntary shifting of usage patterns by customers, thermal energy storage systems, and communication load control.

Peak demand can be flattened through *selective reductions in service* according to procedures prearranged with the customers. Typically, service would be interrupted to cooling and ventilating systems, water heaters, and similar systems in which interruptions can be tolerated for short periods with minimal inconvenience.

Thermal energy storage (TES) systems, an aid to load management, have been demonstrated successfully in residential housing, and also show promise in the commercial sector. Fifty-eight projects for testing TES systems are being conducted in twenty-two states, as indicated in Table 3, which also summarizes several load control projects. Most installations were single-family residences although some were in commercial and public buildings—most notably, a 400,000 square foot warehouse in Toledo, Ohio, and a 268-bed hospital in Manchester, New Hampshire.

The purpose of these projects was to demonstrate load shifting potential, determine customer response, and collect rate-related information. The results are preliminary but indicate that, in properly designed systems, portions of the cooling and heating loads can be shifted to off-peak hours. It is not yet clear whether the increased efficiency of generating during off-peak hours with baseload equipment will offset the energy losses in storage, although similar systems have reduced consumption in on-peak hours in Great Britain during the last few years. Customer response has been generally favorable, but the incentives which must be offered to the U.S. customer to install the TES systems are uncertain. In today's housing market, a typical residential TES system could require an extra initial investment of $300 to $750.

Communication load control is a load management tool designed to permit load shedding during peak hours and otherwise adjust demand patterns. Typically, a signal is sent by the utility to the point of use. The signal is converted into a command which actuates switches controlling the user's demand pattern by turning off some of his equipment

TABLE 3. SUMMARY RESULTS—SURVEY OF UTILITY LOAD MANAGEMENT AND ENERGY CONSERVATION PROJECTS IN THE ELECTRIC UTILITY INDUSTRY

Type of Thermal Energy Storage Project	*Number of Utility Programs*
Water heat storage	4
Pressurized water heat storage	9
Water cool storage	1
Water cool and heat storage	7
Static ice cool storage	14
Dynamic ice cool storage	2
Low temperature ceramic heat storage	2
High temperature ceramic heat storage	5
High temperature ceramic heat storage—central units	4
In-Ground heat storage	5
Ceramic and water heat storage	1
Static ice and water heat and cool storage	1
Annual cycle energy systems	3
Total projects	58

Type of Communication Load Control Project	*Number of Utility Programs*
Radio	14
Power line carrier (high frequency)	10
Ripple (power line carrier—low frequency)	7
Combination radio and power line carrier	2
Time switches	2
Ripple within a single building	1
Time and thermostat switches	2
Thermostat switches	1
Telemetering	1
Telephone	1
Multiple building computer system	1
Total projects	42

Source: Survey of Utility Load Management and Energy Conservation Projects, Part 1, Final Report, December 1977 by Electric Power Research Institute.

for a brief time during periods of peak demand on the utility system. By rolling the control from customer-to-customer, the utility demand can be lowered without greatly impacting the consumer's perceived energy use. Variations include signals to the user (e.g., lights), but most systems eliminate consumer actuation. User acceptance and cooperation have been good; rate incentives have been offered in most projects. These inducements appear to be adequate, but more work is needed to validate the acceptability of communications load control techniques.

Load control systems continue to be used by large-scale industrial users to balance internal system loads and avoid demand charges for peak use. The system may include manual or automatic routines for controlling cooling and ventilating systems in accordance with occupancy patterns, outside weather conditions, and equipment loads. Additional functions can include remote monitoring of heavy electrical equipment so that the equipment can be shut down in the event of incipient failure, malfunction, fire, or other emergency.

Beginning in 1975, the Federal Energy Administration initiated a series of projects to demonstrate and collect usage data on innovative rate designs and load management techniques such as those discussed above. These projects encompassed a wide range of rate structures, with emphasis on the use of time-of-day rates on several systems. The program, which has been continued by the Department of Energy, has included a total of sixteen projects and 18,000 test or control customers.

The Department of Energy is currently performing analysis on data from these projects. Preliminary analysis indicates the following.

1. The amount of energy used by customers on time-of-day rates when compared to that of either control customers or the same customers the year earlier on standard declining-block rates, fell significantly during peak periods and rose during nonpeak periods.
2. For residential customers, the effective reduction in peak period consumption tended to range from 15 to 30 percent in most projects. Figure 6 provides an example of the comparison between the demand by customers with time-of-day rates and a control group (no time-of-day rate). The figure shows the average daily load curve for those two types of customers and indicates that demand during the peak period (12:00 noon to 6:00 P.M.) was substantially reduced by the customers with time-of-day rates, compared to the control group. (Note, however, that the time of peak demand for the time-of-day group was shifted to about 9:00 P.M. Although the peak is lower than that for the control group, the shift time of peak can introduce new problems for utilities.)
3. Analysis of the data on total kWh consumption, however, does not yield a clearly defined set of conclusions. In most projects it was found that the total annual consumption of time-of-day customers was slightly less than that of control customers, by perhaps 5 to 8 percent. However, in at least one project, the electricity consumption of customers on time-of-day rates rose slightly (by about 3 percent).
4. Analysis of data from several projects also suggested that monthly consumption reductions by time-of-day users tended to occur in the summer,

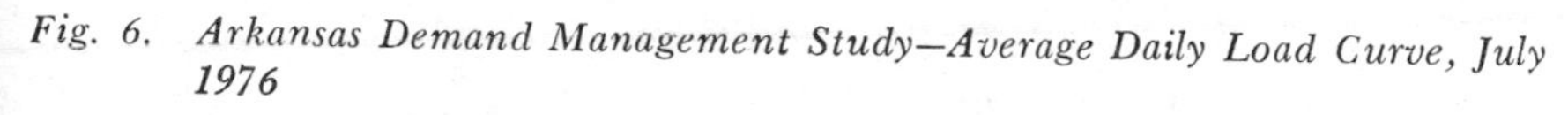

Fig. 6. Arkansas Demand Management Study—Average Daily Load Curve, July 1976

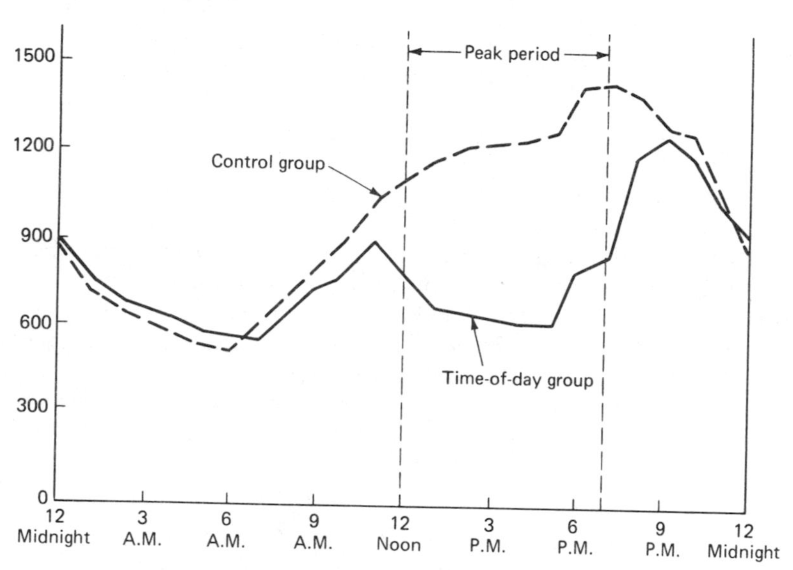

while such users' monthly consumption appeared to be about equal—or slightly above—control customer usage during the winter.

This section has described some of the options which exist to either improve electricity generation efficiency through advances in generation technologies or improve the efficiency of utilization of current capacity through pricing or load management techniques. The next section addresses the institutional barriers to the adoption of efficiency options and related public policy choices.

Institutional Barriers to the Use of Improved Efficiency Options

Given that improvements in conversion and system utilization efficiency can produce gains for consumers and utilities, it may seem surprising that the efficiency improving options have not, as yet, been widely adopted. The causes for the slow adoption derive from the institutional conditions in which utilities and consumers make choices regarding electricity generation and use. Stated simply, conflicting public demands have frustrated the traditional utility decision process and constrained their choice of new conversion technologies. Perceived and real problems with system utilization options have made utilities approach these options with caution. The traditional consumer reluctance to make purchases for energy savings is an additional factor which has slowed adoption of system utilization options. In this section, we will summarize these institutional problems.

For years, electricity demand grew steadily; economies of scale were being achieved through the introduction of larger boilers and improved technologies. Fuel supplies were reliable and prices of primary fuels were steady. In that environment, electric utilities had considerable latitude in their decisions concerning new capacity. Management was able to choose new generation alternatives and obtain Public Utility Commission (PUC) approval usually without difficulty. Since the late 1960s, utility management has been confronted by a radical change in its decision environment. This change had produced institutional barriers to the adoption of approved conversion efficiency options.

Although a detailed description of these institutional impediments is complex, it is possible to summarize, in general terms, their main characteristics. Stated in simplified terms, since the late 1960s, society has placed conflicting and frustrating demands upon utilities. These demands have come from a variety of sources. For example:

1. Environmental regulations have delayed the installation of new capacity. Detailed impact statements must be prepared for each individual site and facility; these often undergo court challenges. The resulting delay produces increased facility costs. Ultimately clean generation of electricity will be more expensive to the consumer as the costs of pollution control equipment are internalized in the rate structure. Pollution control technologies raise yet unresolved implications on power plant availability and fuel utilization.
2. Many PUC's have implicitly demanded that they become instruments of social policy. For example, lifeline rates have been required by these commissions to shelter low-income persons from rapidly rising energy costs.
3. The federal government effort to shift from gas and oil to more coal utilization has focused largely on utilities. Yet, environmental problems associated with coal use have raised the cost of new coal plants by more than 20 percent.
4. Consumers have demanded maintenance of low electricity rates while failing to appreciate the extent to which the incremental cost of new capacity has risen.

Utilities are thus confronted with a variety of seemingly conflicting objectives:

1. They must maintain system reliability yet restrict their use of gas and oil plants while the siting of new coal and nuclear capacity is being delayed by a plethora of new regulations.

2. They must keep the environment clean, while at the same time, increase their utilization of coal and keep their rates down.
3. They must maintain financial viability in the face of regulatory delays and rapidly escalating incremental costs of new capacity.

Public expectations regarding electrical utilities are neither consistent nor stable. The politicization of utility issues creates the problem of attempting to match the twenty- to forty-year utility planning horizon with the two- to four-year political attention span. The skepticism of load management results in a time period consistent with their planning horizon. The consumer resistance to higher rates and in some cases time-varying rates, further deters the use of these options.

Given the large array of new demands, conflicting requirements, and institutional problems, it is difficult to describe *acceptable* public policies which will help to solve the utility dilemma and lead to more efficient electricity use. It is difficult to state acceptable public policy solutions, precisely because of the level of political debate. The society is in the midst of a collision between two different and possibly mutually exclusive paradigms (alternative sets of values concerning social choices which structure the way in which facts and hence reality is viewed). On the one side, politically powerful societal groups are demanding that new generating capacity (among other things) incorporate the full cost of environmental cleanup and protection (including very stringent safety requirements). On the other hand, other groups are pressing utilities to continue their traditional role of providing low cost, reliable power while shifting away from the use of oil and natural gas. The political process, in so far as it is the "art of possible," must be pushed to resolve the results of these conflicting paradigms.

Summary

There are five general conclusions which can reasonably be drawn from this chapter.

1. *There are opportunities for improving the efficiency of electricity production and utilization.* Improvements in the efficiency of *production* depend upon the development of substantially new technologies, e.g., MHD and pressurized fluidized bed systems, rather than refinements and scale-ups of established technologies. Improvements in the efficiency of *utilization* of electricity require much more detailed knowledge of customer usage patterns and their amenability to change under the influence of innovative rate structures and load management technologies that will regulate the time (and efficiency and cost) at which electricity is supplied to users.

As the costs both of fuel and capital equipment have escalated more rapidly than the consumer price index, siting problems have become aggravated, environmental restrictions have increased, and regulatory authorities have proliferated. It has become both economically attractive to seek a better balance between cost-effective substitutions of labor and materials for energy and increased use of energy forms other than electricity.

More attention can usefully be directed toward efforts to improve the *efficiency* of utilization of electricity. Early rate and load management demonstrations provide a basis to predict that these efforts would be fruitful and in the interests of utilities, rate payers, and stockholders.

2. *The problems besetting the electric utility industry are usually system-specific and regional.* Institutional arrangements must be structured so that regional solutions can be accomplished, a difficult task with separate state PUC's and 3,500 individual utilities. Solutions are not likely to be limited by technical inventiveness, but instead may well be limited by a confluence of conflicting public goals and structural barriers to efficiency improvements, i.e., the efficiency of institutions which affect utilities.

3. *Utilities are besieged with a confluence of often conflicting purposes and expectations by the various publics served or affected by their operations.* Amidst the stresses induced by environmental and energy concerns and economic pressures, they still must provide reliable electrical supplies at the lowest reasonable costs to their consumers. To accomplish this underlying purpose, they must undertake parallel activities to keep options open in an uncertain decision environment—activities which inevitably add costs to delivered electricity. There is a clear price associated with a lack of resolution of the nuclear fuel cycle and evolving environmental standards, and, in the end, the public must pay these costs.

4. *In an era of overregulation generally, it is crucial that new regulations placed upon the utility industry be explicit with respect to goals* (e.g., greater efficiency in the use of fuels and capital, greater equity, etc.) but tolerant in the acceptance of different means to accomplish these ends.

5. *National attention must be directed toward the role of electricity vis-à-vis other energy forms in our economy.* Electricity is a versatile and crucial energy form for many purposes but it is being used in applications for which less versatile energy forms might well substitute. To ensure that sufficient reliable electricity supplies are available at prices the public can afford in the future, the scope of electricity applications may well have to be reduced from that scope evident in the present economy. For example, the interactive roles of solar energy (for heating and cooling) and fossil fuels consumed at point of end use must be compared with electricity from several points of view in an economy that is not wholly

"free market" to ensure we are husbanding and directing resources in a way in consonance with the long-term national interest.

While our current debate centers on promoting efficiency to conserve scarce capital and particular fuel forms (e.g., gas and oil), the growing problems with other energy forms strongly suggest a Btu conservation ethic for *all* energy forms and steady attention to implementing changes that would provide the amenities that each of us seek with less energy input and more benign environmental ramifications.

Christopher T. Hill and
Charles M. Overby

8

Improving Energy Productivity through Recovery and Reuse of Wastes

Introduction[1]

The largest industrial consumers of energy in the United States are those industries that extract and harvest raw materials from the earth and process them into useful materials from which things are made. The production of such materials as iron and steel, aluminum, copper, paper, plastics, glass, and cement all require prodigious amounts of energy. As a rule, energy use for production of these materials is significantly greater than the energy required to fabricate final goods from them. This is so, not because the basic materials industries are somehow more wasteful, but because of the technical requirements of the chemical and physical

[1] This chapter is based in part on studies in which the authors participated at the Office of Technology Assessment, U.S. Congress.

CHRISTOPHER T. HILL, *before becoming Senior Research Associate, Center for Policy Alternatives at M.I.T., served in the Office of Technology Assessment, U.S. Congress. Mr. Hill has written extensively in the areas of technology and human affairs.*

CHARLES M. OVERBY, *Professor of Industrial and Systems Engineering, Ohio University, recently spent a sabbatical leave with the Office of Technology Assessment, U.S. Congress. Professor Overby has had a long involvement with technology-society issues.*

operations that must be performed to win useful materials from nature's stocks.

Recovery and recycling of waste materials, as well as reuse and remanufacture of products, have long been practiced as effective ways to conserve materials and, often, to save money. In recent years, it has been recognized that these are also good ways to improve our utilization of energy through avoidance of some of the most energy intensive steps in materials production. Furthermore, waste materials of organic origin such as food and yard waste, paper, wood, and plastics that cannot always be recycled because of technical or economic limitations have come to be viewed as a potential new energy source. In addition to their benefits in terms of materials conservation and improved energy productivity, recycling and reuse also can help to reduce air and water pollution and land disruption associated with production of materials and can help to solve some of the problems associated with ultimate waste disposal.

Despite the large potential benefits of recovery and reuse of wastes, substantial issues have arisen in regard to the technical, economic, institutional, and political feasibility of implementing various strategies. In this chapter, we review the energy savings potential of recovery and reuse of wastes, the status of various technologies and approaches for doing so, and the likely effectiveness of several economic and institutional policies that might be considered for stimulating improved energy productivity through better use of materials. The chapter emphasizes, but is not limited to, resources that are found in, or are destined for, municipal solid waste (MSW).

Some Approaches to Materials Recovery and Reuse

It is useful to think about recovery and reuse within a "materials system" that underlies our economy. Materials can be thought of as flowing through the system from the point of extraction from nature to the point of ultimate disposal. In between, materials are processed and fabricated into products that are used for human purposes. Some materials that become wastes of primary manufacturing and product fabrication are recycled almost immediately. Other materials become parts of products that may be reused, repaired, or remanufactured many times before they are discarded. Still others are fabricated into products that are disposed of after serving only a single brief use.

The materials system is illustrated in a schematic diagram, Figure 1. The central core of the system represents the direct flow of materials from extraction to disposal. The system also includes at least six substantially different modes of materials recycle and reuse, illustrated in Figure 1 as six loops.

Loops 1 and 2 represent the long established industrial practices of

Fig. 1. ***A** Complex Model of the Materials Systems, Showing a Variety of Recycle Loops and Disposal Options*

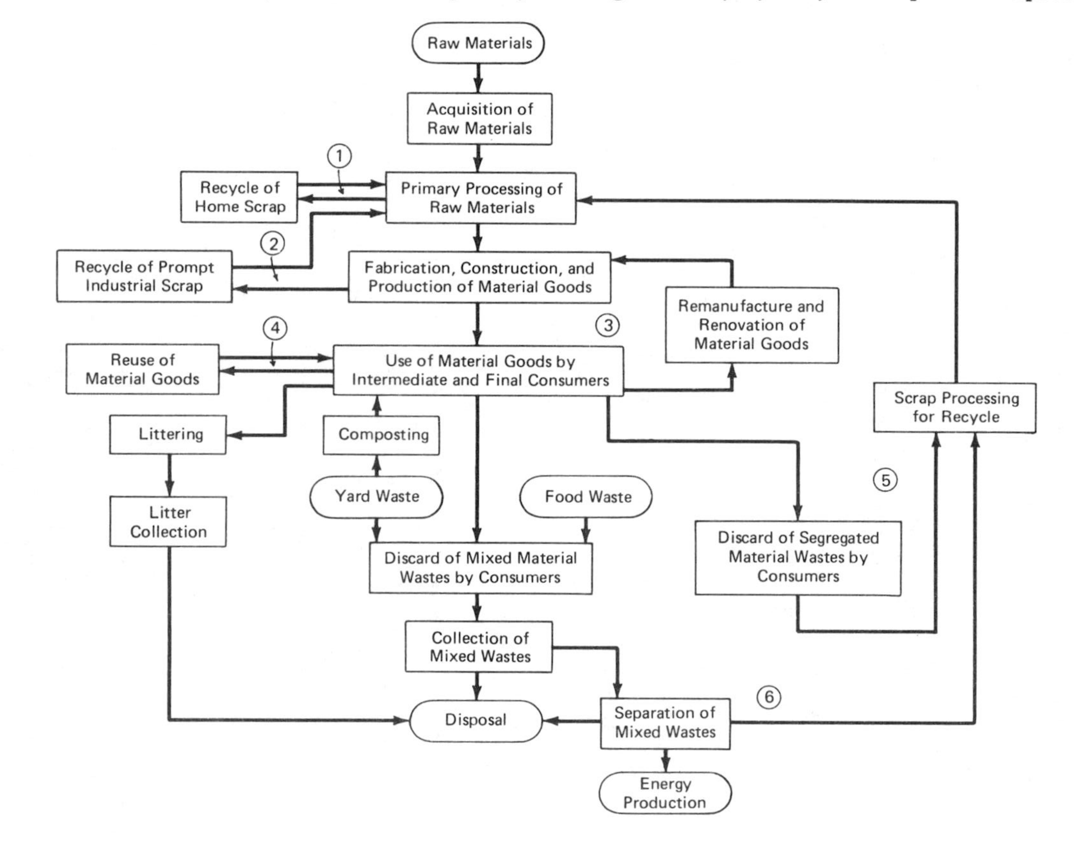

immediate recycle of *home scrap* within a primary materials processing facility or of *prompt industrial scrap* from fabricators directly back to primary processors.

Loop 3 represents a variety of *rework* practices including remanufacture of auto parts, recapping of tires, refurbishing of telephones, renovation of standing buildings, and repair and sale of used clothing and appliances by handicapped workers.

Loop 4 represents direct *reuse* of material goods with little or no change in form. Typical examples of reuse include return of beverage containers for refilling, reuse of "used cars" by second or third owners, and reuse of shipping pallets.

Loop 5 represents *recycling* of discarded material wastes which are *segregated* by material type at each stage in the loop. Several examples of this approach can be cited. Separate collection of one or more components of municipal waste is practiced in a number of areas, often by curbside collection of newspapers, collection of corrugated cardboard at commercial establishments, or "paper drives" sponsored by nonprofit organizations. A second example is "community recycling" in which nonprofit organizations or local governments provide facilities at which citizens can drop off nonmixed wastes such as paper, cans, bottles, and waste oil for recycling. A third example is aluminum can recycling centers operated by aluminum manufacturers and beverage companies. Another example is collection of waste oil from service stations for refining. In each case the segregated wastes can be easily processed because they are kept relatively free of contamination.

Loop 6 represents *recycling of mixed wastes* that are separated to recover materials and fuel that are burned without separation to produce energy. In either case a solid residue for ultimate disposal remains. One example of this kind of recycling is shredding of automobile hulks to remove nonmetals and to produce one or more recyclable metallic fractions. Another example is separation and/or combustion of mixed municipal solid waste in centralized resource recovery plants. Such plants are being developed to recover various recyclable materials (such as ferrous metals, aluminum, glass, and mixed nonferrous metals) as well as to produce such energy products as refuse-derived fuel, steam, electricity, pyrolytic gas or oil, or biologically produced methane gas.

Figure 1 also shows yards waste being returned to users. This can be done by individuals at home or by collection, composting, and redistribution of such waste as compost and mulch, as is practiced in some communities.

Contrary to popular impressions, America is already a reusing and recycling nation. For example, nearly two-thirds of the nation's iron and steel output is produced from recycled home, prompt, and postconsumer scrap. The scrap iron and steel industry had sales of $3 billion in 1977.

Likewise, substantial fractions of the nation's copper, lead, and construction paper production are based on scrap as raw materials. There are flourishing industries that remanufacture items ranging from auto parts and aircraft engines to copying machines and computers.

The current boom in renovation of urban residences is another manifestation of reuse of obsolete products. On the other hand, there are clearly areas in which recycling and recovery could be expanded, not only to save materials and energy, but to preserve the value-added embodied in products that are prematurely disposed of.

Perhaps the most obvious (but not necessarily the most productive) area for improved recovery and recycling is municipal solid waste. The United States generates annually more than 135 million tons of MSW. Disposal of this waste is a rapidly growing problem in many areas of the country, which are finding such traditional disposal methods as open dumping, landfill, uncontrolled incineration, and ocean burial too expensive or environmentally unacceptable. At the same time, this MSW contains over half of the national consumption of paper, two-thirds of the glass, over one-fifth of the aluminum, and nearly one-eighth of the iron and steel. If burned, this waste could supply somewhat more than 1 percent of the nation's energy needs. Today, however, only about 6 percent of this MSW is recovered and recycled.

Another area in which more might be done is the remanufacture, refurbishing, and recycling of consumer durables such as appliances, furniture, and electronic equipment. These activities are inhibited by the great variety of brands and models that results from the use of product differentiation as a marketing strategy for such products, by rapid technological advance in some of them, and by the fact that consumer durables are not designed to be easily disassembled or repaired.

The remainder of this chapter examines in more detail the opportunities and problems associated with resource recovery from municipal solid waste and with reuse and remanufacture of consumer and industrial products. In each case, government policies that might be implemented to enhance recovery and reuse are discussed, and the energy implications of such policies are examined.

Energy Savings through Materials Recycle

THE CONCEPTS

Recycling, as used in this chapter, means the reprocessing of waste materials and products to produce new basic materials. One example of recycling is melting of scrap iron and steel from discarded auto hulks, "tin" cans, and railroad rails to produce new basic steel shapes such as reinforcing bars, flat plates, or tubes. Another example is repulping of waste newspaper to produce new newsprint. Yet another example is melting broken, used glass containers to produce new beer bottles.

As compared to production from virgin raw materials, production of basic materials by recycling saves energy by avoiding the primary processing and transportation of ores, fluxes, pulpwood, and the like. Thus, the production of raw steel by melting scrap in an electric furnace avoids the energy used in mining and transporting iron ore, coal, and limestone as well as the energy used in the primary chemical reduction of iron ore to pig iron in the blast furnace. In this case, the percentage of energy savings due to recycling are large. Similar conclusions hold for other metals such as copper, aluminum, lead, and zinc. On the other hand, since production of glass involves no energy consuming chemical reactions and only one melting step, substitution of scrap glass for virgin sand and soda ash has a smaller potential for energy savings.

It is more difficult to assign energy savings to recycling of waste paper. In the first place, much of our paper today is produced from forest industry wastes that might otherwise be discarded or used as fuel themselves. Second, the virgin paper industry gets much of its energy input from forest wastes that do not appear in the nation's fossil energy budget, whereas paper recycling mills use more fossil fuels as inputs. Third, waste paper can itself be used either as a scrap raw material or as an energy source. As a result of these complexities, the proper accounting for energy productivity in recycling paper has yet to be agreed on by analysts, policy-makers, environmentalists, and the industry.

The energy implications of plastics recycling are also complex. There are two main classes of plastics, the thermoplastics such as polyethylene, polystyrene, and nylon, and the thermosets such as epoxies, polyesters, and phenolics. The thermoplastics can be recycled by remelting and processing into new products, although with some loss of quality upon each remelting. This remelting operation requires only a small fraction of the energy required to make the plastic originally. However, unlike mixed steels, mixed thermoplastics do not alloy well, if at all, and their melting temperatures are not high enough to burn off the organic contaminants that can be removed from metals this way. Therefore, successful remelting requires that plastics be separated by chemical type and that all contamination be removed before melting. Since separation and decontamination of thermoplastics are difficult and expensive tasks, only home and prompt scrap are currently recycled. Postconsumer thermoplastics may, therefore, contribute the most to energy productivity by being burned directly or with mixed wastes as fuel.

The thermoset plastics cannot be remelted at all. Due to their chemical structure, rather than melting at elevated temperatures, they break down into various organic gases, liquids, and solids. Experiments have been performed to use these degradation products as new chemical feedstocks, but it may be economically and energetically preferable to use waste thermosets, which will burn, as a fuel. Alternatively, they can be mechanically broken up and used as a lightweight aggregate or filler in concrete, asphalt, or new thermosets.

DATA ON ENERGY SAVINGS THROUGH MATERIALS RECYCLE

In recent years, a number of studies have been made of the energy savings potential of materials recycling. In doing such studies, one first ascertains the amount of energy required to produce one unit of a basic material (say, one pound of steel or of polyethylene). This is done by adding up all the energy inputs to that unit including process and transportation fuels, mining, raw materials processing, heat and light for operation of facilities, and so on. Energy input data are based on the performance of actual or average "model" plants. One also calculates the energy required to recycle used materials to produce the same unit of basic material. The difference between the two energy values is the energy savings made possible by recycling.

Typical data on production energies for virgin and recycled basic materials are shown in Table 1. The data illustrate the points made above. Recycling of aluminum, copper, and thermoplastic polyethylene plastic can save a large fraction of the energy required to produce these materials from virgin raw materials. Recycling of steel is somewhat less effective in saving energy, and recycling of newsprint and of glass saves even less energy.

TABLE 1. ENERGY USE IN PRODUCTION OF MATERIALS FROM VIRGIN AND SECONDARY RAW MATERIALS

	Production Energy (Btu/pound)		
Material	*Virgin Inputs*	*Secondary Inputs*	*Secondary Energy as a Percentage of Virgin Energy*
Steel	8,300	7,500 (40% scrap)	90
		4,400 (100% scrap)	53
Aluminum	134,700	5,000	3.7
Ingot	108,600	2,200–3,400	2.0–3.1
Copper	25,900	1,400–2,900	5.4–11.2
Glass containers	7,800	7,200	92
Plastics (polyethylene)	49,500	1,350	2.7
Newsprint	11,400	8,800	77

Table 1 illustrates another important point; the energy required to produce one pound of various materials differs widely—compare steel and aluminum. Keeping in mind that different amounts of different materials are required to perform specific tasks, it is clear that the energy productivity of materials use can be influenced by choices among materials as well as by a choice between virgin and recycled materials. This life-cycle energy productivity aspect of materials choice has been extensively examined. One finding of such analyses is that from a life cycle energy point of view, aluminum and plastics are often preferred to steel in automobiles and other vehicles because their higher production energy is more than offset by their lighter weight and consequent reduced transport fuel requirement.

The reader should be aware that there are considerable uncertainties in the numbers quoted in Table 1. These uncertainties arise from the practical difficulties in accounting for all energy inputs to materials production as well as from fundamental conceptual limitations regarding the mode of accounting to be used and the treatment of energy inputs of different thermodynamic quality. Thus Table 1 is indicative, rather than definitive.

Energy Savings through Recovery of Resources in MSW

POTENTIAL SAVINGS

As noted earlier, MSW can be viewed as a source of both recyclable materials and recoverable fuels. Table 2 shows estimates of the amount

TABLE 2. POTENTIAL ANNUAL ENERGY SAVINGS FROM RECOVERY AND RECYCLING OF ALL OF VARIOUS COMPONENTS OF MSW

Material or Fuel Form	*Energy Savings (10^{15} Btu/year)*
Materials	
Iron and steel	0.08
Aluminum	0.19
Copper	0.01
Glass	0.02
Newspaper	0.03
Total potential energy savings by materials recycle	0.33
Fuel value of combustibles except newspaper *	1.17
Total energy savings and fuels produced	1.50
Total energy savings and fuels produced as a percent of annual national energy use	2.1%

* Fuel value of newspaper in MSW is about 0.13×10^{15} Btu.

of energy that could be saved if all of the various materials and potential fuels in MSW could be recovered and recycled. In this computation, newspaper (and other paper) can be viewed as a potential fuel or as a recyclable material. Plastics, wood, textiles, rubber, food and yard wastes, and nonnewspaper have all been treated as combustibles rather than recyclables.

Table 2 shows that about 2 percent of the nation's energy use could be met or avoided by burning or recycling the materials contained in municipal solid waste. Most of these "savings" could come from burning the combustible fraction; the remainder from energy saved through recycling metals, glass, and newspapers.

However, it is not at all clear that the potential savings suggested by Table 2 are technically, economically, or politically feasible. The next several paragraphs describe the status of two major approaches to recovery of materials and energy from waste: source separation and centralized resource recovery. Policies directed at implementation of these two approaches are also discussed.

STATUS OF SOURCE SEPARATION FOR MATERIALS RECOVERY

Source separation (collection of unmixed MSW at curbside, at community recycling centers, or in offices and institutions) is a viable way to recover high quality recyclable glass, aluminum, and ferrous metal and is the only way to recover waste paper that is suitable for recycling to new paper. Source separation is also an attractive way to collect organic wastes for composting.

Several hundred source separation programs of various types are in operation around the United States. These include municipal collection of newspaper and, occasionally, glass and cans; collection of high quality office paper waste and computer cards; industry sponsored programs for aluminum can collection; and community drop-off centers for paper, glass, and cans. No comprehensive estimate has been made of the nationwide total amount of materials recovered by these methods. However, nearly all of the 7.9 million tons of MSW recycled in 1975 was recovered in source separation programs. This amount is equivalent to 6 percent of the total MSW generated. Today source separation programs provide the raw materials for rapidly growing industries that recycle newsprint, corrugated cardboard, glass bottles, and aluminum cans.

The key factor in the success of a source separation program is the participation rate, which is defined as the percentage of each source-separable component of the waste that is actually recovered. If a high 75 percent participation rate were achieved in a comprehensive source separation program, we estimate that 41 percent by weight of an average community's waste might be recovered for recycling. One-third of this amount would be yard waste for composting, two fifths would be paper, one-fifth would be glass, and the remainder would be steel and aluminum cans. A residue of 59 percent of the waste would still be available as an energy source, or would need to be disposed of by other means. The degree of recovery would be proportionally less at lower participation rates. For example, at 25 percent participation, only 14 percent of the waste would be recycled and 86 percent would remain.

Source separation programs can be implemented rapidly and with relatively modest capital investment. For communities that are too small to support a centralized resource recovery facility, source separation may be the only way to recover valuable materials. In contrast to mechanical separation of collected mixed wastes in centralized resource recovery plants, source separation requires a high level of cooperation by generators of waste. For this reason and because separate collection can require expansion of expensive collection activities, successful source separation programs require considerable attention to design and implementation strategies.

It has proven difficult to determine whether source separation programs are cost-effective. Revenues and landfill credits can be estimated easily from amounts recoverable and realizable prices. Costs are more elusive, because they are a part of the costs of collection of waste, and these costs are not easily allocated among collection functions. Data from programs in Andover and Marblehead, Massachusetts, and Madison, Wisconsin, suggest that curbside residential pickup programs can be expected to break even (and to make a net profit when waste newspaper prices are high).

From the state and local perspective, source separation programs would be stimulated by any federal policy that strengthens the demand for recovered materials or by any federal environmental policy that makes landfill less attractive or more expensive than it has been. Either of these approaches would make source separation, as well as other resource recovery methods, more attractive economically.

Beyond economic and environmental programs, three aspects of federal policy toward resource recovery are important for source separation programs. First, federal technical assistance and educational programs under the Environmental Protection Agency (EPA) and the Department of Energy can treat source separation as a meaningful approach to resource recovery. Second, federal support could be increased for demonstration programs designed to show how source separation might be implemented in large urban areas, in order to move beyond the affluent suburban image that is often associated with source separation. Third, modest federal support could be provided for research and development (R&D) on devices, procedures, and systems for separate collection and for upgrading of collected materials to meet user specifications for quality.

STATUS OF CENTRALIZED RESOURCE RECOVERY FOR MATERIALS AND ENERGY FROM MSW

A number of technologies exist in various stages of development for the centralized recovery of resources (energy and materials) from MSW. In this section we report on the status of these systems and touch on several related policy issues.

Centralized resource recovery is a rapidly emerging industry. In their 1977 Fourth Report to the Congress, the Environmental Protection Agency listed 118 facilities that were in various stages of development and planning in the summer of 1976. Twenty-one facilities were in an operational status, ten were under construction, thirty-three were in advanced planning, and fifty-four communities were sufficiently interested to have either commissioned or carried out informal feasibility studies. EPA identified only seventy comparable facilities in 1974. There is a tendency toward large-scale plants with about half of those under construction or in advanced planning in 1976 having a capacity of 1,000 tons per day or more.

Performance of Alternative Technologies—The variety of resource recovery technologies is listed in Tables 3 and 4 according to their degree of commercialization. This classification scheme, while of a judgmental nature, is an aid in assessing the near-term feasibility of energy and materials recovery on a commercial basis for any specific system or for any specific material. Thus we see from Table 3 that the only commercially proven energy recovery systems are waterwall combustion, small-scale modular incineration with heat recovery, and solid refuse derived fuel (RDF). Commercially proven materials recovery technologies (Table 4) are humus from composting, magnetic recovery of ferrous metals, and

TABLE 3. DEGREE OF COMMERCIALIZATION OF ENERGY RECOVERY TECHNOLOGIES

Commercially operational technologies
Waterwell combustion
Small-scale modular incineration with heat recovery
Solid fuel RDF (wet and dry processes)
Developmental technologies
Low Btu gas pyrolysis
Medium Btu gas pyrolysis
Liquid pyrolysis
Methane recovery from landfills
Experimental technologies
Biological anaerobic digestion
Waste-fired gas turbine
Research technologies
Hydrolysis

TABLE 4. DEGREE OF COMMERCIALIZATION OF MATERIALS RECOVERY TECHNOLOGIES

Commercially operational technologies
Composting
Magnetic recovery of ferrous metals
Fiber recovery by wet separation
Developmental technologies
Aluminum recovery
Glass recovery
Experimental technologies
Nonferrous recovery
Paper recovery by dry processes

recovery of low quality fiber with the wet processing system. Technologies for recovery of aluminum and glass, for pyrolysis to produce gases and liquids, and for recovery of methane gas from landfills are currently under intensive study and are being installed in commercial sized plants. At least some of these are expected to become commercially operational in the early 1980s.

There is currently no standardized and accepted way to compare the energy recovery efficiencies of energy recovery technologies. In their Fourth Report, the EPA reported energy recovery efficiencies that range from a high of 63 percent for "dust RDF" to a low of 23 percent for liquid fuel pyrolysis, based on conversion of waste to steam. Estimates such as these must be viewed with caution since they are based on engineering estimates and not on actual operating systems. They also ignore important economic characteristics of waste-derived fuels such as the quality of the fuel product or steam and its transportability.

Estimates of the efficiency with which materials can be recovered in centralized systems are given in Table 5. Except for iron and steel, these

TABLE 5. MATERIALS RECOVERY EFFICIENCIES

Technology	*Percent of the Waste Stream Content of Each Material Recovered*
Iron and steel	90–97
Paper fiber–dry	23
Paper fiber–wet	50
Aluminum	65
Glass-froth flotation	65–70
Glass-optical sorting	50

TABLE 6. ESTIMATED WASTE REDUCTION EFFICIENCIES OF RESOURCE RECOVERY TECHNOLOGIES

	Residue as Percent of Input Waste	
Technology	*Weight Percent*	*Volume Percent*
Waterwall combustion	20–35	5–15
Small-scale incineration	30	10
Dry fluff RFD	10–20	–
Low Btu pyrolysis	15–20	3–5
Medium Btu pyrolysis	17	2
Liquid pyrolysis	7	1–2
Anaerobic digestion	17	–

estimates are based on experimental or developmental plants, and they may not reflect actual performance in day-to-day operation.

The waste reduction capability of centralized resource recovery systems is another factor of great interest to municipal decision-makers. Table 6 gives literature estimates of the degree to which the various technologies reduce the weight and volume of the residue that must be disposed of after processing.

Substantial uncertainties remain concerning the nature and severity of potential environmental and occupational health problems associated with centralized recovery. Three areas are of particular concern: (1) pathogens in the plant environment, (2) heavy metals and complex hydrocarbons in air emissions from combustion or pyrolysis, and (3) potential water pollution from landfill of the residual solids from incineration and pyrolysis. Accelerated federal efforts are needed to identify and define these hazards, to develop control technologies or process modifications to manage them, and to set regulatory standards for their control. The current state of affairs not only risks unknown threats to public health but also poses a barrier to implementation of resource recovery because of the uncertain future regulatory environment.

R&D efforts are needed on a variety of hardware and systems questions related to centralized resource recovery. However, private firms are quite active in developing and promoting new technology. Consequently, the federal role in R&D might best be confined to basic research on separation and utilization of waste and to research on the nature and control of environmental and occupational health hazards of these technologies.

This approach to R&D would reflect the priorities and strategies suggested by Gibbons in Chapter 10.

Economics of Centralized Recovery of Materials and Energy from Waste—The economics of centralized resource recovery are complicated because they involve capital-intensive, unproven technologies that enjoy considerable, but uncertain, economies of scale. On the benefit side are revenues from recovered materials and energy, credits for avoided landfill, and credits for reduced social costs of waste disposal by other means. On the cost side are capital and operating cost, additional costs of transfer and transportation of MSW to a distant, central facility, and the loss of flexibility in waste disposal implied by a capital-intensive approach.

In today's markets, energy provides around two-thirds or more of the potential revenues from centralized resource recovery, while materials provide one-third or less. One ton of incoming MSW might yield energy revenues of six to fifteen dollars, materials revenues of three to six dollars, and landfill avoidance credits of two to ten dollars, for total revenues and credits of eleven to thirty-one dollars per ton. Thus, considerable attention must be paid to the matching of resource recovery plants with energy consumers, particularly since energy transportation costs are high. Since very large energy consumers such as electric utilities have not proven to be strong markets for recovered energy, it may prove advantageous to focus on smaller resource recovery plants that can serve the more numerous smaller energy consumers such as office buildings, shopping centers, educational institutions, and small factories. Additional federal agency effort is needed to define the nature of potential consumers of recovered energy and to design programs that will be responsive to the actual market situation.

Despite the importance of energy revenues to the economics of centralized resource recovery, resource recovery is not cost-effective as an energy supply option alone. It is economically feasible as an energy source only because it can take credit as a disposal method as well.

For example, total costs of building and operating a centralized facility, including extra costs for transportation, are likely to be in the range of twenty to thirty dollars per ton. As noted above, total energy and materials revenues and landfill avoidance credits are likely to be in the range of eleven to thirty-one dollars per ton, minus charges for delivery of recovered products to users. Thus, centralized resource recovery can be expected to break even only where energy prices and landfill costs are high. The fact that these conditions prevail there helps to explain why current interest in resource recovery is centered in the urban parts of the northeastern United States.

Costs of landfill are expected to increase in the next several years as convenient sites are filled and as standards of environmental control under the Resource Conservation and Recovery Act of 1976 are implemented and enforced. This fact alone is expected to make recovery of energy and materials from MSW more attractive.

It is difficult to project the impact of rising energy prices on the economic viability of resource recovery, largely because its capital costs, which are a large part of the total costs, can be expected to rise with the energy prices. (See Pindyck's Figure 3 in Chapter 2, which shows that real prices for capital and labor increased together in the 1970–74 period.) Thus, if landfill credits were excluded from the balance sheet, resource recovery might not become economically self-sustaining unless energy prices were to increase much more rapidly than capital costs. With landfill credits included, resource recovery can become economically feasible in larger and larger fractions of the nation as energy prices increase. A future doubling of real prices of energy relative to capital prices would probably make resource recovery economic in nearly all urbanized parts of the country.

According to some economic modeling studies, large, centralized resource recovery plants with significant economies of scale are economically feasible today in some heavily urbanized areas, if local institutional problems can be overcome. Thus, direct subsidy programs in these circumstances would be both unnecessary and expensive. The direct federal role there can be limited to education, technical assistance, and planning aid.

Not enough is known about the economics of small-scale incineration technologies to determine whether they would be economically self-supporting in smaller communities without subsidy. It is known, however, that local jurisdictional conflicts and resistance to regionalized approaches are real and major barriers to regionalized, multicommunity resource recovery.

Institutional Barriers to Resource Recovery—There are a variety of institutional barriers and problems in the implementation of resource recovery by both separate collection and centralized separation. Some of these problems may be more difficult to solve than the technological and economic problems discussed above. In fact, in the context of resource recovery, recycling, and reuse, some institutional problems may be essentially insurmountable; the only option may be to circumvent them by adapting technology to existing institutions, by adopting new economic incentives or disincentives, or by establishing entirely new institutions.

Table 7 lists the most important institutional problems that have been identified for centralized resource recovery. Space does not permit discussion of the details of these problems or of the potential of federal

TABLE 7. INSTITUTIONAL PROBLEMS IN CENTRALIZED RESOURCE RECOVERY

Inadequate or unavailable information

1. Technological uncertainty
2. Economic uncertainty
3. Inadequately informed local citizens, technical staff, and decision-makers
4. Oversell of technology by vendors and proponents

Jurisdictional problems

5. Fragmented and overlapping state and local jurisdictions
6. Cost sharing among jurisdictions
7. Overlapping federal agency jurisdictions
8. Responsibility for and title to waste after discard: "flow control"
9. Limitations on interjurisdictional waste shipment or disposal

Implementation problems

10. Siting of facilities
11. Citizen acceptance
12. Construction delays
13. Limited bonding capability of local governments
14. Inadequate health, safety, and environmental standards for resource recovery plants
15. Cooperation of local waste collectors and haulers
16. Creation of local monopolies in centralized resource recovery

Marketing Problems

17. Inadequate or nonexisting standards of performance for recovered products
18. Fluctuating scrap prices and demand
19. Limited ability of local governments to enter into long-term sales contracts
20. Electric utility rate regulation which discourages use of new fuel sources

actions to resolve them. Clearly, public agencies or private entrepreneurs who wish to explore resource recovery face a formidable array of difficulties that transcend simple cost-effectiveness considerations.

Table 8 lists federal actions that have been or might be undertaken to help overcome the institutional barriers and problems. It is convenient to divide these options into direct federal actions and indirect federal actions, implemented through requirements for state and local actions under federally supported programs. A number of these options have been adopted under the Resource Conservation and Recovery Act of 1976 or under other legislation.

The federal government obviously has available a wide variety of options for addressing institutional problems in resource recovery. However, caution is warranted in adopting these policies, since they may resolve some problems, while exacerbating others.

TABLE 8. FEDERAL POLICY OPTIONS TOWARD INSTITUTIONAL PROBLEMS IN CENTRALIZED RESOURCE RECOVERY

Direct federal actions

- Construction subsidies: grants, tax credits, low interest loans, loan guarantees *
- Operating subsidies: recycling allowance, direct product subsidy, labor tax credit
- Research, development, and demonstration funding *
- Education and training *
- Technical assistance *
- Planning grants *
- Mandated product standards
- Mandated federal procurement of recovered materials *
- Freight rate adjustment or subsidy *
- Stockpile for recovered materials
- Issue regulations for environmental, health, and safety performance of resource recovery facilities
- Interagency coordination *

Indirect federal action through requirements for state and local actions under federally supported programs

- Mandate regional solid waste planning * and operation
- Encourage private sector involvement
- Encourage or mandate opportunities for citizen participation *
- Mandate reform or electric utility ratemaking framework
- Remove barriers to local long-term contracting for solid waste disposal, resource recovery, and product sale *
- Encourage utility commissions to regulate resource recovery firms
- Prohibit or repeal "flow control" ordinances which mandate use of particular resource recovery facilities
- Encourage facilitated site selection and permitting
- Mandate cost-sharing formulas for local jurisdictions
- Extend right of eminent domain to resource recovery systems
- Mandate state and local procurement of recycled materials and recovered energy when available

* Option implemented by current law, executive order, or agency decision.

Some Policies to Stimulate Reuse and Remanufacture of Products

BEVERAGE CONTAINER DEPOSIT LEGISLATION (BCDL)

One policy option for reducing the rate of generation of solid waste through reuse of consumer products is federal legislation to mandate a minimum, refundable deposit for all containers used in the sale of carbonated alcoholic (beer) and nonalcoholic (soft drink) beverages. Some proposals would also mandate deposits on containers for mineral water, wine, spirits, milk, fruit juice, iced tea, and related beverages.

In its simplest form, deposit legislation would require that all parties in the distribution train, from brewers and bottlers to retailers, charge a minimum deposit, say five cents per container, which is refundable upon presentation of an empty equivalent container. While such legislation would neither mandate use of returnable containers nor ban use of nonreturnable ones, the intent of BCDL is that containers be used that are either refillable after cleaning or recyclable into new containers or other goods.

Under BCDL, various studies have estimated savings in total energy use for beverage delivery ranging from 21 to 61 percent, with estimates clustered around 40 percent savings. Had such a law been in effect in 1975, savings of 40 percent would have meant savings of 170 trillion Btu's per year or the equivalent of 80 thousand barrels of oil per day. These estimates take into account all energy use for beverage delivery including container production, transportation, storage, retailing, collection, washing, and recycling. If beverage prices were to decline under BCDL, some of the energy savings might be offset by consumer spending of the money saved on beverage purchases.

Consumption of aluminum, ferrous metal, and glass for beverage containers under BCDL would be very dependent upon the final market shares and return rates for each container type. Generally, for any reasonable set of market shares and return/recycle rates, aluminum and steel consumption would decline. Glass consumption would also decline, if return rates were greater than 80 percent, even if an all-glass returnable system were to result.

Total nationwide generation of MSW has been projected to decline by 1 to 5 percent under BCDL, with the actual value being very sensitive to the returnable bottle return rate. BCDL might reduce the materials content of MSW by as much as 37 percent for aluminum, 36 percent for glass, and 11 percent for ferrous metal, but not all of these reductions would occur together.

Based on observations in Oregon and Vermont, which have implemented state deposit laws, highway beverage container litter has been projected to decline under a national deposit law by around 80 percent. Total highway litter might decline by 20 percent on an item count basis. However, these estimates are highly uncertain.

A deposit law would cause an increase in capital investment in brewing, beer wholesaling, soft drink bottling, and retailing. Output and required investment in can and bottle making would decrease. Total costs of doing business would increase in wholesaling and retailing and would decrease in bottling and brewing. Total costs of beverage delivery and beverage shelf prices might increase or decrease slightly, depending largely on the competitive responses of the industries involved. Shelf price changes of –4 to +9 percent have been suggested. Deposits not reclaimed by some consumers make an important contribution to offsetting additional costs and should not be subject to special tax or confiscation.

Employment of skilled workers in can and bottle production might decrease sharply under BCDL, with reductions being especially severe in communities where these industries are major employers. Employment of workers in beverage production, delivery, and retailing would increase, with increases being distributed rather uniformly across the country. The net effect of BCDL would be to cause an increase in the total number of workers employed and in total wages paid.

Federal programs would be desirable to ameliorate the undesirable side effects of adoption of BCDL, which would fall heaviest on container producers and workers. Various kinds of assistance might be provided to involved industries and workers. This assistance could take the form of retraining and relocation allowances, accelerated depreciation, or plant conversion assistance for container producers. For beverage delivery industries, policies might be adopted to share the business risk associated with the transition. However, care must be taken to avoid subsidization of activities, such as equipment modernization, that might have been undertaken by firms even without BCDL. A reasonable transition period might reduce the disruptive effects of BCDL, but a period of more than two to three years would probably be counterproductive because firms would delay their responses as long as possible.

Beverage container deposit legislation is only one of several possibl product reuse options. One might consider mandatory deposits on a hos of reusable goods such as automobiles, tires, electrical machinery, con sumer durables, or food packaging, some of which could be reused an others of which would require some degree of remanufacture.

REMANUFACTURE OF PRODUCTS

Energy and materials embodied in obsolete products can be saved b remanufacturing, refurbishing, or repairing them for further productiv use. In this section, a conceptual framework for remanufacturing is estab lished, potential energy savings in a typical operation (tire retreading are estimated, and possible government policies for stimulating remanu facture are identified. Remanufacture has only recently captured the in terest of analysts, so the data base for this discussion is very thin.

In remanufacturing operations, a reasonably large quantity of simila products is brought to a central facility and disassembled. Parts of specific product are not kept together. Instead, they are collected by par type, cleaned, and inspected for possible repair and reuse. Products ar then reassembled, usually on an assembly line basis, using recovered parts and new parts where necessary.

In refurbishing, products are also returned to a central facility for processing. However, upon disassembly, a product's component parts are kept together and after cleaning, inspecting, and replacing with new parts where necessary, the original product is reassembled from most of its original parts. Refurbishing is not so easily arranged for mass production methods as is remanufacturing.

Repairing includes those situations in which individual products are returned to a functional state, one at a time. The process involves troubleshooting to identify the malfunction and then appropriate disassembly, repair, and reassembly of a single product, as in television repair shops, automobile garages, and "do-it-yourself" repair by consumers.

Automobile tire recapping provides an interesting example of energy savings through product remanufacture. Today, the equivalent of approximately 20 percent of all vehicle tires produced annually are retreaded. This amounts to some 45 million tires, whose mileage lifetimes can range up to 90 percent of those of a new tire. As an upper limit, contrasted with a situation of no recapping, we have estimated the energy that could be saved if all passenger car tires were to be retreaded on the average of one time before being discarded. Using some crude approximations, it would appear that about 0.065×10^{15} Btu of energy could be saved each year (about 0.1 percent of the nation's energy consumption) if all tires were retreaded once.

There would, of course, be additional impacts of the retreading program just described. First, the output of the new tire industry would drop by nearly half, with a concomitant decrease in employment. Likewise, the demand for synthetic rubber for tires would decline by around one-fourth, again with employment losses. On the other hand, employment and business opportunities in tire recapping would increase, while tire disposal problems would be cut by half. The net impact on business and jobs is not at all clear at this point. However, a shift in jobs from large companies (tire producers) to smaller ones (handlers, retreaders) would be expected, with an associated shift of employment regionally. Retreading is probably more labor intensive than tire making, so total employment might increase. The price of tires, at least to second owners, would be less than the price of new ones.

This example illustrates the thorny distributional questions associated with remanufacturing. Generally, one can expect shifts from large, centralized firms to smaller, less centralized ones. Further, jobs are likely to shift from well organized, highly skilled, but highly differentiated ones, to less organized, lower skilled, but more diverse ones. Similarly, decentralization of the geographic location of manufacturing facilities and jobs is expected. On the other hand, a trend toward leasing, with original equipment manufacturers retaining ownership, could lead to concentration and different forms of centralization. At the moment, however, very little is known about the nature and magnitude of these kinds of effects.

There are other important implications of remanufacturing. Among these are possible tendencies toward greater use of materials and energy in new products that are designed to facilitate remanufacture, potential trends toward standardization and reduced innovation in products that might experience more rapid innovation if sales volumes were higher, and potential complications of liability for injury caused by remanufactured products.

A wide variety of federal policies could be considered to stimulate remanufacturing, refurbishing, and repair of products. These include deposits and bounties to encourage return of used products; tax or other incentives and disincentives to adjust the competitiveness of new and remanufactured products; requirements for federal procurement of remanufactured products; a variety of informational programs including education, training, and technical assistance; and regulation of remanufacturing firms or licensing of repair persons. A great deal of work remains to be done in assessing the effectiveness and impacts of such policies in the areas of energy, materials, employment, investment, innovation, and so on.

Economic Policies to Stimulate Recycling

Many proposals have been made over the years for government to stimulate recycling through various incentive and disincentive programs. Unlike earlier sections that focus on MSW, this section applies to recycling more generally. In this section we examine the potential contributions of six such proposals.

1. The product disposal charge
2. The recycling allowance or subsidy
3. The severance tax
4. Modification of the percentage depletion allowance for minerals
5. Modification of the capital gains treatment of timber income
6. Adjustment of railroad freight rates for virgin and scrap materials

ECONOMICS OF MATERIALS DEMAND

Before examining the effects of these proposals on recycling, it is useful to review some general economic principles that govern the response of flows in the materials system to changes in material prices, whatever their cause. This section is an elaboration for the materials system of the economic principles set forth by Pindyck in Chapter 2.

Evaluation of the response of material flows to economic policies is based on the principle that the rate of consumption of a material is influenced by its price, by the associated costs of using it, and by the prices and costs of using alternatives to it. Five general responses might follow a change in the relative prices of materials. Suppose that, as a result of a new government policy, a recycled material were to become available at a price relative to the price of its virgin material counterpart which is lower than previously had been the case. Furthermore, assume that all other prices and costs in the economy were to remain fixed. Any or all of the following outcomes might occur:

1. Increased output from some industries that use the recycled material as an input.
2. Substitution of the recycled material for the virgin material in some applications.
3. Substitution of the recycled material for some other material in certain applications.
4. Substitution of the recycled material for other factors of production such as labor or capital in some applications.
5. Development of new technologies or emergence of new industries that use the recycled material as an input.

The times required for these five responses to occur are very different—they are listed above in order of increasing time scale. The first three responses are said to occur in the "short run," which may be several days to a year or two. For example, output increases may occur in a few weeks if cost savings are passed on to customers and they increase their purchases as a result. Substitution of recycled for virgin materials may also be rapid if existing equipment can be used with either one. Similarly, in some industries recycled material of one kind may be easily and rapidly substituted for some other material to make a product. "Long-run" responses (those of the fourth type) may only occur after several months or several years, and usually involve changes in capital equipment and in the work force in order to use more of the recycled material and less capital or labor. The fifth kind of response to price change is technological innovation, which usually occurs only in the "very long run" and may take from a year or two up to ten or more years to develop.

The ability of analysts to predict the magnitudes of the five kinds of responses listed above is better developed for the short run than for the long run. Most studies of the response of demand to price are based on short-run models that only capture responses of the first three kinds. The widely used input-output methodology for estimating the energy impact of changes in final demand can only encompass responses of the first kind because it is based on the assumptions that technologies and ratios of capital, labor, and materials are constant over time. The short-run responses are also the only ones captured in most elasticity analyses, as noted by Pindyck in Chapter 2. Only in recent years have analysts paid attention to economy-wide responses of types two, three, and four; and studies of technological innovation (type five response) in response to materials prices have only begun to appear. One study found that six or more years are required for price changes to cause significant process innovation in production of anhydrous ammonia. As a result of these analytic shortcomings, most studies of the response of material flows to changes in their prices tend to underestimate the changes in material flows that would ultimately occur.

In addition to limits on analytical methods of prediction, there appear to be limits on the intuitive perceptions of observers about the nature and magnitude of the response of demand to price changes, especially for basic industrial inputs like materials and energy. Perceptions tend to be based on short-run experience, and such experience teaches, for example, that if steel prices go up, the auto industry continues to sell just as many cars and to buy just as much steel; hence, demand for steel appears not to be responsive to price. Yet this is only the type one response. In the long run, in response to higher steel prices, auto companies may use lighter gauge metal, adopt less wasteful machining techniques, shift to aluminum, or build smaller cars. Since other things also change in the long run, sorting out the part of any change that has occurred in response to material price changes may defy the best "intuiters," not to mention the most careful analysts.

Intuition is even less reliable as a guide to anticipation of the response of demand to changes in prices of scrap materials. Historically, the short-run demand for a secondary material usually increases, rather than decreases, when its price increases. This phenomenon is caused by the fact that when the short-run demand for scrap on the part of producers of primary materials is high due to strong demand for goods in general, scrap dealers can charge higher prices. The short-run variations in scrap prices tend to be greater than their long-run movements, and this also masks our ability to observe and understand long-run behavior.

The "availability" of materials also tends to influence their relative flows in the economy; all other things being equal, users will tend to select the material that is more available. Availability is related to price response, but is less well-defined. In the short run, a material is perceived to be "available" if the supply is highly responsive to price, that is, if purchasers can buy all they need at the normal price or slightly above. If this is not the case, or is perceived not to be the case, the material is said to be less available. Such short-run availability is closely related to the fact that productive capacity may not be easily or quickly adjusted. In the long run, availability is related to physical exhaustion of the resource base, or in the case of scrap, to exhaustion of the available scrap inventory. Political factors also affect perceived availability. For example, the existence or possibility of new environmental restrictions, labor actions, or international market disruptions may adversely affect availability.

EFFECTIVENESS OF POLICIES FOR AFFECTING MATERIALS USE

With these limitations on our ability to analyze the response of use and recycling of materials to prices, availability, and economic policies, let us return to our examination of the six policy options. The options are defined as follows:

The Product Charge—an excise tax levied on material goods proportional to their weight, volume, or other measure of disposal cost. The tax would be levied on material fabricators or related industries. An exemption to the charge is offered for use of recycled materials.

The Recycling Allowance—a direct grant or tax incentive to producers or users of recycled materials paid in proportion to some measure of the amount or value of recycled materials used.

The Severance Tax—a tax on virgin materials levied at the point of mining or harvest in proportion to some measure of the amount or value extracted.

The Percentage Depletion Allowance—existing law allows for deduction from income before taxes each year of a percentage of gross income from mining specified minerals. For the purposes of this analysis, modification or repeal of this deduction is examined.

Capital Gains Treatment of Income from Standing Timber—existing law allows for taxation of income from sale of standing timber at rates appropriate to long-term capital gains, which are lower than rates for ordinary income. For the purposes of this analysis, modification or repeal of this tax preference is examined.

Adjustment of Railroad Freight Rates—it has been widely argued that rail rates discriminate against recovered materials. Adjustments of such rates might assist marketing of recycled materials.

The product charge and recycling allowance are specifically designed to encourage recycling and discourage waste of materials. The severance tax has traditionally been used by states as a revenue measure, rather than as a recycling incentive. The percentage depletion allowance for minerals and capital gains treatment of income from standing timber are tax preferences designed to aid specific industries; recycling was not originally a factor in establishing either of these policies.

For some of these policies, quantitative assessments of effectiveness are available from the literature. The product charge, which might result in a 2 to 10 percent reduction in MSW generation and in substantial increases in recycling, might be most effective in the near term. However, its effectiveness would be strongly dependent on successful administration of an exemption to the charge for recovered materials, and there are many potential problems in implementing the exemption.

Various forms of the recycling allowance would be somewhat less successful than the product charge and would have similar administrative shortcomings.

Repeal of the percentage depletion allowance on minerals or repeal of the capital gains treatment of timber income would increase recycling by only a fraction of a percent to a few percent. In addition, these actions are not expected to significantly reduce the generation of waste, although quantitative estimates of this impact have not been made. Furthermore, percentage depletion and capital gains treatment of timber income are important factors in the cost structure of the affected industries, and any modifications to achieve small increases in recycling may not be warranted. Quantitative estimates of the effectiveness of the severance tax have not been made, but it would be expected to be about the same as repeal of percentage depletion or of capital gains on timber.

In discussion of the freight rate issue, three different definitions of discrimination that follow from different approaches to rate-making have been used. Whether discrimination exists, as well as the nature of that discrimination, depends on the definition used. Using the marginal cost pricing definition as a basis for freight rate adjustment (which gives the largest and most consistent indicator of discrimination against scrap), only small percentage changes in scrap shipments are likely for ferrous metal, waste paper, and aluminum. Waste glass recycling rates might increase by 11 to 17 percent. An unknown portion of these increases would come from MSW.

Discussion and Conclusions

This chapter has been concerned with showing some of the ways in which recovery and reuse of products, materials, and energy from waste might contribute to improving the nation's energy productivity. Direct connections between materials recycling, waste recovery, and energy productivity were illustrated, and less direct but important connections were established between energy utilization and reuse and remanufacture of products such as beverage containers and automobile tires. Finally, some evaluations of the effectiveness of economic policies in stimulating recycling and, therefore, energy savings were reviewed.

In this brief treatment, many subtleties, much detail, and many connections to energy productivity have had to be omitted.

Much work remains to be done in assessing the effectiveness and implications of the use of materials policy to improve energy productivity.

It is not always true that using less material means using less energ In fact, designs that make buildings, vehicles, and energy conversio equipment more energy efficient may require use of more materials or use of materials whose production is more energy intensive. In the area of remanufacture and repair, our understanding of technology, economics, and policy has only begun to emerge.

Denis A. Hayes

9

hort-term Solar Prospects

We are *not* running out of energy. Ninety-nine percent of all the energy that will ever be available for human use is in the sunlight that strikes the earth. Humankind's current commercial energy budget equals only 0.01 percent of this solar influx; a hundredfold increase would equal only 1 percent. Fundamental physical limits will constrain energy growth long before it multiplies a hundredfold.

We are not depleting this most plentiful source of energy. Unlike fossil and fissile fuels, sunlight is a flow and not a stock. Once a gallon of oil is burned, it is gone forever; but the sun will continue to cast its rays earthward, whether sunshine is harnessed today for human needs or not. Technical improvements in the extraction of finite fuels could hasten their exhaustion; similar improvements in the use of sunlight could lower prices permanently. The sun will eventually collapse upon itself, becoming a dense "black hole" that emits no energy at all. But the best guess of informed scholars is that this event will not occur for three or four billion years.

Although we are not running out of solar energy, we *are* running out of cheap oil and gas. The United States was largely built on oil that cost two dollars per barrel. International oil now costs $13.50, and some analysts expect this price to rise to twenty-five dollars per barrel within the next ten years. As the price of our most important energy source climbs steadily higher, dramatic social changes may be unavoidable.

Most such changes will manifest themselves in the substitution of relatively cheaper factors of production for energy. Materials (e.g., insu-

---

DENIS A. HAYES *is Senior Researcher with Worldwatch Institute, a private organization in Washington which analyzes global issues. Formerly he was director of the Illinois State Energy Office.*

lation and storm windows) will be substituted for energy. Capital
a more energy efficient, but more expensive, aluminum refining proce
will be substituted for energy. Intelligence (e.g., replacing much ma
delivery with telecommunications) will be substituted for energy. And labor (e.g., a mechanic tuning up a car or extra retail clerks needed to handle returnable bottles) will be substituted for energy.

Such substitutions will allow finite supplies of scarce fuels to be stretched much farther than would otherwise be the case—a necessary first step requires the gradual substitution of plentiful energy sources for ones in more limited supply. At the heart of the Nixon, Ford, and Carter energy plans has been a policy of substituting coal and nuclear power for oil and natural gas. To a much lesser extent, these plans have also envisioned the eventual substitution of solar-based renewable energy resources for finite fuels.

A substantial body of opinion within the energy policy community holds that the United States will eventually obtain virtually all its commercial energy from the sun. Many believe the solar transition can begin today. The President's Council on Environmental Quality (CEQ) estimates that a combination of renewable energy resources could contribute 25 percent of the nation's energy budget by the year 2000.

If the long-term energy future of the United States is to be built on a renewable base, and if renewable resources can make a substantial contribution within twenty-two years, large current investments in coal and nuclear facilities are arguably a mistake. Renewable energy technologies are (with very few exceptions) best applied in a decentralized fashion. It makes little sense to convert from our present decentralized energy system (based on oil and natural gas) to a highly centralized system (built around coal and nuclear power) as a transitional step toward a decentralized solar-powered society.

Can solar technologies achieve widespread acceptance fast enough to make the CEQ target possible? I suspect the CEQ goal may, in fact, be too modest. The proposition is not amenable to "proof" one way or the other. But it might be helpful to examine some analogies and then explore the physical requirements for such rapid growth.

Rapid Growth Is as American as Television

Americans are capable of extraordinary exertions in times of perceived national crisis. The mobilization following Pearl Harbor and the launching of the first Soviet space satellite are often cited as models for rapid solar development. The energy crisis has not yet inspired the necessary public will for such an all-out effort. But another dramatic interruption in the flow of foreign oil, or any major disaster at a nuclear facility, might evoke a sea change in public consciousness.

ıericans are also given to widespread individual responses to ideas ι strike their fancy. Fifty percent of all U.S. families, for example, ɔw have home gardens—up from about 20 percent in 1973. An estimated 40 percent of Vermont homeowners retrofit their homes with wood-burning stoves in three years. Sales of bicycles, C.B. radios, skateboards, and pocket calculators over the last few years would have absolutely confounded anyone trying to project them using the techniques ordinarily employed in energy policy analysis.

One of the most dramatic examples of the proliferation of a new technology was the rapid saturation of the national market for television. In 1946, about five thousand Americans had television sets. By 1956, the total had shot up to 42 million. By 1966, the number had grown to 65 million, which is about where it has remained. The market went from essentially zero to saturation in twenty years.

In 1946, no one could have guessed the explosive growth to come. The elasticity data on those first five thousand television sets was as useless as the assumptions drawn from today's five thousand solar-heated homes. They were expensive items with obvious public appeal but unproven marketability.

Some will find such "reasoning by anecdote" amusing. Sales of television sets, jeeps, and pocket calculators are interesting, but they really do not have much to do with serious business of energy policy. Currently, solar energy sources contribute only 6 percent of the nation's commercial energy—mostly as hydropower and combustible wastes of the forest products industry. Bemused skeptics "know" that it is flat out impossible for renewable resources to capture an additional 20 percent of the total U.S. energy market in the next twenty-two years.

Yet such a development would not be unprecedented. In the twenty-two years between 1949 and 1971, natural gas captured 20 percent of the national energy market, growing from 18 percent to 38 percent. Environmentally clean and easily employed in a decentralized fashion, natural gas became the fuel of choice for tens of millions of individual families and businesses. Governmental subsidies and price controls make it economically attractive as well. If natural gas could achieve such a growth rate during a calm, business-as-usual period in U.S. history, it does not seem unreasonable to believe that renewable energy sources could now match its success if backed with a major federal commitment during a period of perceived national crisis.

Let us examine a bit more closely just what a 25 percent goal would entail. Obviously, the first question to answer is "25 percent of what?" The most impressive effort to date to forecast U.S. energy demand, reported in *Science,* April 14, 1978, produced a range for the year 2010 spanning from 58 to 136 quadrillion Btu's—depending upon energy prices and policies. Let us assume, conservatively, a range from 85 quads to 125 quads in the year 2000. The question then becomes, "can some combination of solar sources displace 21 to 31 quads of conventional primary fuels by the year 2000?"

We will reserve for the moment the question of whether such a program makes sense in terms of economics, environmental quality, national security, intergenerational equity, and practical politics. The desirability of a program—by any criteria—is of little significance if the goal itself is physically impossible to attain. Hence, we will examine the physical dimensions of one possible combination of solar technologies that could significantly exceed the 25 percent goal and allow the reader to judge the plausibility of such a strategy. Then, if satisfied that we are discussing a credible option, we will assess the desirability (on economic and other grounds) of pursuing it and evaluate some of the major policies that might be useful in such a pursuit.

Residential Solar Heating

Heating accounts for a greater fraction of national energy demand than is commonly appreciated. About 58 percent of all energy end use in the United States is in the form of heat. Most of this (35 percent of all energy) is at temperatures less than 100° C—achievable using simple, inexpensive solar collectors.

Perhaps the simplest solar device is the residential water heater. The collector is, in essence, a box with a black bottom and a glass or plastic top. Glass is transparent to sunlight but not to radiation of longer wavelengths given off by the black collector bottom. Hence, heat is trapped inside. When water is pumped through the hot collector, its temperature rises. The hot water is then piped to a well-insulated storage tank where it is kept until needed. In 1941, Florida had 60,000 solar water heaters—more than the entire nation has today. As early as 1897, 30 percent of the homes in Pasadena, California, had solar water heaters. Two hundred thousand homes in Israel and two million Japanese homes have solar water heaters today.

Sunshine can also be used to heat buildings. "Passive" systems store energy right where sunlight strikes the building's walls and floor. Such systems are designed to shield the structure from unwanted summer heat while capturing and retaining the sun's warmth during the colder months. Passive solar architecture is, beyond doubt, the most efficient and cost-effective way to heat and cool new buildings, and well-designed passive solar greenhouses are often an inexpensive way to retrofit existing structures. Comparatively modest investments will often provide 80 percent or more of a new building's space conditioning requirements. "active" systems, fans or pumps move solar-heated air or liquid f collectors to storage areas, from which heat is withdrawn as needed.

self-sufficiency will often dictate a combination of active and passive features.

In the year 2000, the United States will have over 100 million dwellings, of which about 45 million are yet to be built. In 1978, in some parts of the Southwest, one quarter of all new homes employed significant elements of passive solar design. Let us assume that one-eighth of the existing housing stock can be retrofit with solar equipment (mostly water heaters and passive solar greenhouses) that provides one-third of their space and water heating. Let us further assume that one-half of all homes built between 1980 and 2000 obtain 75 percent of their heat from passive and active solar features. If the retrofits are assumed to all replace primary fuels (i.e., oil and gas) and the new construction to replace efficient electric heat pumps, about 7.5 quads of primary fuel will be displaced.

Commercial and Industrial Solar Heating

Commercial floor space is expected to more than double during the next twenty-two years, with about one-fifth of current floor space disappearing. Solar features incorporated in new structures can rather easily meet 80 percent of the space and water heating requirements of many commercial buildings. Passive space heating systems can also be added to a large fraction of the existing stock of commercial buildings; generally this requires little more than covering the southern wall with an outer sheath of glass. Let us assume that one-tenth of all commercial buildings obtain 80 percent of their heat from the sun and that one-fifth of all commercial buildings obtain 50 percent of their heat using solar equipment. If half of these solar savings displaces primary fuels and half displaces electric heat pumps, about 5.5 quads of primary fuel will be displaced.

Solar heating technologies have industrial applications as well. A study of the Australian food processing factory, for example, found that heat comprised 90 percent of the industry's energy needs. Almost all this heat was at under 150° C and 80 percent was below the boiling point of water. Such low-temperature heat can be easily produced and stored using simple solar devices. In the United States, solar heating is already being applied to a soup canning plant in California, a fabric drying facility in Alabama, and a concrete block factory in Pennsylvania. Solar-powered laundries and car washes are now operating in California, and a beer brewery in Jacksonville, Florida, has turned to solar heating. The Inter-Technology Corporation, under contract to ERDA, estimates that 7.5 uads of such industrial heat can be provided by solar devices by 2000 mostly in the chemical, textile, and food processing industries. Let us ume that about one-fourth of this—2 quads—will be tapped. Commer- and industrial applications combined will then yield 7.5 quads.

Wind Power

The air that envelopes the earth functions as a vast storage battery for solar energy. Winds are generated by the uneven heating of our spinning planet's land and water, plains and mountains, equatorial regions and poles. The windmill played an important role in American history, especially in the Great Plains where it was used to pump water. More than six million windmills were built in the United States over the last century, and about 150 thousand still spin productively. Prior to the large-scale federal commitment to rural electrification in the 1930s and 1940s, windmills supplied much of rural America with its only source of electricity.

There is some controversy over the size of the possible energy harvest by land-based wind turbines in the United States. The major uncertainty is over just where these machines can be erected. Most estimates range between one to two trillion kilowatt-hours per year—or roughly ten to twenty quads. The Bureau of Reclamation recommends the establishment of "wind farms" linked to hydroelectric facilities. The wind turbines would generate power when the wind blows, and water turbines would take up the slack whenever the wind dies down. Dams have sufficient excess storage capacity in the seventeen western states to accommodate a 2.5 quad tandem wind harvest.

Proposed commercial wind turbines range from a few kilowatts to 2.5 megawatts, and some combination of small and large will doubtless prove optimal. Let us assume we build up our production and maintenance capacities to the point where by 1985 we can begin installing 13 thousand megawatts of wind turbine capacity per year. (If this output were all in one megawatt machines, the amount of material processed each year would equal about 2 percent that of the U.S. automobile industry.) By the year 2000, the cumulative installed capacity would be 200 gigawatts. Operating at a capacity factor of 30 percent, these wind turbines would displace five quads of primary fuel in the year 2000.

Photovoltaic Cells

Solar cells generate electricity directly when sunlight falls on them. They have no moving parts, consume no fuel, produce no pollution, operate at environmental temperatures, have long lifetimes, require little maintenance, and can be fashioned from silicon—the second most abundant element in the earth's crust—as well as from many other common materials.

Photovoltaic cells are modular by nature, and little is to be gained by grouping large masses of cells at a simple collection site. On the contrary, the technology is most sensibly applied in a decentralized fashion—e.g., the roofs of buildings—so that transmission requirements can be minimized. With decentralized use, the 80 to 90 percent of the energy in sunlight that is not converted to sunlight can be harnessed for space heating and water heating. The U.S. Department of Energy's goal is to make photovoltaics cost-competitive for grid applications (i.e., for the toughest market) by 1986. A large body of evidence suggests that this goal—fifty cents per peak watt—can be achieved with a strong federal program.

Assuming photovoltaic cells that are 15 percent efficient, 13 billion square feet would yield about 200 peak gigawatts of generating capacity. If produced between 1985 and 2000, this goal would require an average annual production of about one billion square feet—roughly equal to the current output of the U.S. glass industry. It would cover about one-fifth of the residential roof area in the country, or about one-third of the residential and commercial roof area to be constructed during the remainder of this century. Distributed throughout all parts of the United States, this installed capacity should produce three quads per year.

Hydropower

Hydropower is currently the largest source of solar-derived electricity. It is the dominant source of power in the Northwest and Southeast. Hydroelectric facilities currently contribute three quads in an average year, and dams now under construction will boost that by 0.2 quads. Many small hydroelectric facilities have been phased out because other generating facilities were less expensive; 112 such sites have been abandoned in New England since 1941—many of which are economically viable today. About 14 thousand megawatts is available at unfilled generation bays in existing large dams, and many other large dams could be retrofit with more efficient turbines. Many smaller dams, mostly built for flood control, irrigation, and recreational purposes, could be economically altered (e.g., with a bypass channel) to generate power. Upgrading all existing hydropower facilities and adding generating capacity to appropriate small facilities would yield 54,600 megawatts of additional capacity. This incremental capacity would produce 2.5 quads of energy per year. If we assume that 1.8 quads of this is tapped by the year 2000, hydropower will contribute five quads *without requiring the construction of any new dams.*

Bioenergy

Biological energy sources fall into two broad categories: waste from nonenergy processes (such as food and paper production) and crops grown explicitly for their energy value. Since waste disposal is unavoidable and often costly, converting waste into fuels—the first option—is a sensible alternative to using valuable land for garbage dumps. However, the task of waste collection and disposal usually falls to those who cling to the bottom rungs of the economic and social ladder and, until recently, waste seldom attracted either the interest of the well-educated or the investment dollars of the well-heeled. But change is afoot, partly because solid waste is now often viewed as a source of abundant high-grade fuel that is close to major energy markets.

The second plant-energy option, the production of "energy crops," will probably be limited to marginal lands, since worldwide population pressures are already relentlessly pushing food producers onto lands ill-suited to conventional agriculture. Yet much potential energy cropland does exist in areas where food production cannot be sustained. Some prime agricultural land could also be employed during the off-season to grow energy crops.

The most familiar energy crop, of course, is firewood. A good fuel tree has a high annual yield when densely planted, resprouts from cut stumps (coppices), thrives with only short rotation periods, and is generally hardy. Favored species for fuel trees are eucalyptus, sycamore, and poplar; an intelligently planned tree plantation would probably use a mixture of species.

Trees are not the only energy crop worth considering. A number of other land and water crops have their advocates among bioconversion specialists. Land plants with potential as energy sources include sugar cane, cassava (manioc), sunflowers, some sorghums, kenaf, and forage grasses, as well as other plants (notably of the genus *Euphorbia*) whose sap contains a rich emulsion of hydrocarbons.

Biological sources currently contribute about two quads of energy—mostly as residues of the forest products industries. This contribution can be increased to about three quads simply by applying the best current practice to the entire industry.

Aggregated nationally, municipal solid waste contains about 1.2 quads of energy. A study for the Ford Foundation Energy Policy Project placed an upper limit on recoverable energy from crop residues, feedlot manure, and urban refuse of just over four quads. The volume of such wastes is commonly expected to double by the turn of the century. It does not seem overly ambitious to tap three quads from these sources in the year 2000.

It should be possible to achieve a net yield of two additional quads from agricultural and silvacultural crops raised explicitly for their energy value. These would be mostly converted into liquid and gaseous fuels. If all these things were achieved, bioenergy sources would displace eight quads of conventional primary fuels. This number is an agreement with an estimate by the Solar Working Group of the CONAES Committee of the National Academy of Sciences.

None of these proposals seems intuitively to be far-fetched. Yet collectively, they would yield thirty-six quads—five quads more than needed to produce 25 percent of the highest of our energy budgets for 2000, and more than 40 percent of the national energy budget I would consider most likely.

Moreover, this list of technologies is by no means exhaustive. For example, it excludes "power towers"—the principal beneficiaries of federal research support to date. These centralized power plants are thought by some to be capable of contributing 1.4 quads by the year 2000, mostly through "repowering" existing conventional power plants in the Southwest. Also omitted are Ocean Thermal Energy Conversion facilities, thought by their proponents to be capable of producing up to forty quads annually. The CONAES Solar Research Group estimates the maximum OTEC contribution in 2000 at 1.6 quads. Small heat engines can be harnessed to solar sources to generate electricity, run irrigation pumps, and perform other work. Evaluations by Battelle Memorial Institute, the Jet Propulsion Laboratory, and the Congressional Office of Technology Assessment all conclude that, with mass production, these would provide one of the least expensive sources of electricity. Theodore Taylor, in a comprehensive evaluation of solar options, concluded that a combination of solar ponds and heat engines offered the brightest promise for lowering costs.

Many of these technologies can be combined in total energy systems for greater overall efficiency. By 1981 a textile factory employing 150 people in Shenandoah, Georgia, will use a solar total energy system to produce electricity, process steam, heat, and cool. Similar combinations of technologies can be employed to provide energy for communities.

The goal of deriving 25 percent of the national energy budget from renewable sources by the year 2000 seems unlikely to be constrained by the resource base or by technology. Having established that it is possible, let us now examine whether it is desirable.

Solar Costs

Conventional wisdom holds that while solar energy has many attractive characteristics, it is too expensive today to see widespread application. As is so often the case with conventional wisdom, yesterday's truth has become today's misapprehension. Parts of the following analysis were suggested by M. A. Maidique in "Solar America" in *Coming to Terms: The Energy Crisis as it Really Is.*

One reason that solar energy is considered uneconomical is that it is commonly judged in terms of a "payback" period. If, for example, a top-of-the-line solar water heater is professionally installed, it might cost $2400 and save $200 a year in oil bills. This leads to a twelve-year payback of the initial investment or an 8 percent rate-of-return. That is roughly the yield from a good industrial bond. The calculation is simple and easily explained to a customer; it is also inadequate and misleading.

For example, the 8 percent yield is not *income* but *savings*. It represents after-tax income the consumer would otherwise spend on fuel. How much gross (pretax) income it represents will depend upon the consumer's tax bracket. Someone in a 33 percent tax bracket would have to obtain a 12 percent gross return to net the same after-tax income; for someone in the 50 percent bracket, the yield is equal to a 16 percent return on a taxable investment.

Moreover, the preceding calculations assume that the price of oil will rise no faster than the general rate of inflation. They assume that next year's savings, and those of subsequent years, when discounted, equal 200 current dollars per annum. More realistically, let us assume that oil prices rise 2 percent faster than inflation. Moreover, let us assume we can obtain a loan that spreads the cost of the initial investment over a twenty-year period. This analytic approach, known as life cycle costing, yields a *tax-free* 15 percent rate of return equal to a gross return of more than 20 percent to someone in the 33 percent tax bracket.

The economic benefits to society from a solar investment are even clearer than the benefits to an individual investor. Society wants to obtain as much energy as possible per dollar of investment. The question, then, is which *new* investment produces the most bang for the buck.

This question is never addressed by the homeowner trying to decide whether to purchase some solar equipment. He compares the price of the solar collector with the average price of fuel. Oil prices, for example, are an average of cheap oil from old domestic fields, imported oil, oil from Alaska and other remote places, and advanced recovery oil obtained by using sophisticated techniques to pull still more oil out of wells that have stopped producing. The cost of just "new" oil is much higher than this average. Some would argue that the real replacement cost should be the cost of making oil from coal—thirty to forty dollars per barrel.

The individual consumer, who has to shoulder the full cost of a new solar water heater, never confronts the cost of new oil. Instead the price of new oil is mixed with the price of cheaper old oil, and all consumers pay for the expensive new oil with slightly higher bills. This effect is most dramatic for electricity. In some parts of the country, power from a new power plant costs ten times more than the average price (including power from cheap dams, etc.) now paid by the consumer.

If everyone paid for oil at the world price, for gas at the price of imports, and for electricity at roughly the cost of power from the least expensive new power plants, the nation's annual energy bill would be $70 billion higher. This market distortion amounts to almost one billion dollars per year per quad of commercial energy sold. If society subsidized solar technologies to this same degree, the impact would be revolutionary. If by the year 2000, solar technologies produced thirty-six quads, solar consumers would be the beneficiaries of a $34 billion annual subsidy.

Throughout the continental forty-eight United States, solar house heating now makes economic sense at the margin. That is to say, if the new energy is to come from an unsubsidized new nuclear power plant or an unsubsidized new solar unit, the solar investment will be cheaper. If society's scarce capital is to be invested efficiently, public policies must ensure that the microeconomic interests of individual consumers coincide more closely with the macroeconomic interests of the nation.

Nonetheless, as we have seen, even where the homeowner must compare the marginal costs of new solar equipment with the average cost of competing energy sources, solar investments will generally make sense over the lifetime of the building. The most important first step is to incorporate passive solar design into the building's blueprints. Often this costs little or nothing. For example, it costs no more to place most windows in the southern wall than to place them facing north, but southern windows capture the sun's warmth while northern windows merely leak the building's internal heat. Roof overhangs will let in the low winter sunshine while protecting the room from a hot summer sun riding higher in the sky. Masonry floors and working shutters are not expensive. Yet, combined with tight construction and good insulation, they can lower the heating load of the building by 17 percent and more. In Arkansas, 200 well-designed houses constructed under a grant from HUD cost no more than neighboring houses built using conventional construction standards, but their fuel bills are only one-fourth as high.

Skeptics often point to the high-cost solar equipment used in some federal solar demonstration projects and claim that the buildings could have been heated with oil for a fraction of the cost. But such buildings incorporate no cheap passive solar features, and they generally employ the most expensive solar hardware on the market. Moreover, the price of oil is an average price—not the price of new oil—and this average price is kept artificially low through depletion allowances, intangible drilling cost write-offs, foreign tax credits, unsafe tankers built for accelerated depreciation, polluting refineries in the Caribbean, etc. Direct federal subsidies to conventional fuels run in the neighborhood of $8 billion per year, dwarfing the comparatively trifling subsidies for solar sources. Even in this loaded contest, solar energy compares fairly well.

Wind turbines currently on the market range from $205 per peak kilowatt to over $1,000. Even in the higher price range, wind turbines (with a 30 percent capacity factor) are roughly competitive with new nuclear power plants (with a 55 percent capacity factor). As wind machines become increasingly mass-produced and as large turbines become commercially available, costs should concentrate toward the lower end of the range. At these prices, wind power is one of the cheapest sources of electricity available, handicapped only by the need for a backup power source or a very large grid to supply electricity during calm periods.

Photovoltaic cells, at $6,000 per peak kilowatt, are not now competitive for grid applications. Until prices fall to under $1,000 per peak kilowatt, photovoltaics will make economic sense only in remote applications and in grid-less developing countries. However, the U.S. Department of Energy has a 1986 goal of $500 per peak kilowatt, and a consensus in the commercial photovoltaic industry holds that this goal will be met or surpassed.

Dams are rather expensive, but retrofitting turbines into existing dams is a relatively inexpensive approach to providing base-load power. The costs for developing the U.S. potential included in this paper would range from $300 to $1,200 per kilowatt.

Biological energy sources are diverse, and it is difficult to accurately assess the potential costs of some approaches. Since municipal waste must be disposed of in any case, generally at a cost of two to ten dollars per ton (but occasionally as high as twenty-five dollars per ton), technologies that convert waste into fuel will generally receive a credit. With such a credit, energy recovery from municipal solid waste is economically viable today (see Chapter 8). The simplest such technology, waterwall incinerators that generate steam for district heating or to produce electricity, is widely employed in Europe. More advanced approaches, yielding a variety of outputs, will be more costly. In general, technologies to tap the energy in "energy crops" will probably require somewhat less capital investment and somewhat higher fuel costs than facilities to produce synthetic fuels from coal. When environmental externalities, including the intractable CO_2 problems associated with coal combustion, are considered, bioenergy sources should be clearly competitive.

The most impressive obstacles to the rapid development of renewable energy resources are not technical or economic; they are social and institutional. To be sure, some are quite formidable. Yet none appears to be insurmountable, as long as solar maintains its very broad base of public support.

This broad base of support does not rest on the macroeconomic factors we have been considering. Indeed, although polls indicate that 94 percent of the public favors a program of aggressive development for renewable energy resources, most people continue to think that the technologies to tap these resources remain "uneconomical." How then is the popularity of solar energy to be explained?

Part of the answer doubtless lies in the simple fact that although most policy-makers are economists, most people are not. It is extraordinary how the tools of the economics profession—and the implicit values embodied in those techniques—have come to dominate the public policy arena. Although "energy economists" were few and far between in 1973 (most earned their livings by giving elaborate explanations of why OPEC would never amount to anything), today Washington is filled with the breed. The fraction of authors in this book who build their cases solely upon economic analysis is indicative of the paramount role now enjoyed by this rather narrow (and notoriously imprecise) discipline. It is, perhaps, enough to remark that if any energy facility enjoyed a 25 percent price advantage over a competitor, there would be no question in the minds of most policy analysts which one to choose; no other criteria would be relevant where there was such a great price gap.

Yet there is abroad in the land a popular sentiment that believes that a 25 percent premium might not be an outrageous price to pay for an energy source that produces neither radioactive wastes nor bomb-grade materials, that neither destroys the landscape nor threatens to trigger a planetary greenhouse effect. Renewable energy resources, used wisely, will have fewer and less serious environmental impacts than conventional energy resources. Renewable resources tend to foster decentralization, pluralism, self-reliance, and a favorable balance-of-payments. They tend to be labor intensive, stable, and resilient. Moreover, used internationally, they should tend to discourage nuclear proliferation. Such elements should be of great value to society. Viewed from the vantage of our great-grandchildren a hundred years hence, such characteristics will almost certainly be of far greater significance than even a 25 percent price difference. Although almost totally ignored in the process of policy analysis, these noneconomic features provide a compelling rationale for establishing public incentives to encourage solar investments.

Public Policies

A substantial body of research has documented the barriers to rapid solar development, and a number of policy initiatives have been designed to overcome these obstacles. While there does not exist much hard data on how effective various policies and programs would be, it might be instructive to quickly review some of the major alternatives.

At the top of most lists are substantial tax credits for investments in renewable resources. Although after a delay of eighteen months, Congress passed a tax credit modeled upon the Carter administration's proposal. The credit is too modest (20 to 30 percent, compared with a 55 percent state credit in California); it applies to too narrow a range of equipment (mostly active solar heating systems); it expires too quickly (1985); it is not "refundable" (hence is of little or no value to people who pay little or no federal tax); and the industrial credit is too small to compensate for the fact that companies can write off expenditures on conventional fuels directly against their taxes as operating expenses.

An interesting case can be made that tax credits should be available not just to final consumers but also to lessors and other middlemen. These parties would then lease or rent solar equipment to apartment dwellers or to businesses seeking to "expense" their solar bills.

Prospective solar purchasers should have access to long-term financing at low rates of interest. At present, homeowners who seek financing are mostly confined to five-year loans at rates of interest in the 15 to 18 percent range. Businesses have also experienced difficulty financing solar investments. One option would be a federal solar development bank to provide a secondary market for loans to solar purchasers and to guarantee loans to solar manufacturers.

An alternative, and in some ways more attractive, way to guarantee very attractive financing in the residential solar market is to employ latent leverage in the existing secondary market. The Government National Mortgage Association, the Federal National Mortgage Association, and the Federal Home Loan Mortgage Corporation could require most national lending institutions to adopt the program currently being used by San Diego Federal Savings and Loan—the nation's fifth most profitable multibillion dollar savings and loan.

San Diego Federal will provide thirty-year financing to any existing mortgage holder who wishes to purchase solar equipment. The existing mortgage remains at its original interest rate; the increment (covering 100 percent of the cost of the solar equipment) is financed at the currently prevailing mortgage rate. The heart of the program is the fact that the homeowners' monthly payments do not change when the solar loan is obtained. Rather, the mortgage is rewritten over a longer period of time, so that instead of being paid off in, say, twenty-two years, it is paid off in twenty-six years. When the solar equipment is installed, the homeowner pays lower average conventional energy bills, pays the same mortgage bills, and obtains a substantial tax credit from the state government. Few homeowners retain ownership for the entire duration of the mortgage; the full mortgage is paid off when the home is sold.

Several renewable energy technologies suffer high costs because they are produced in low volumes by cottage industries. While safeguarding against the capture of the solar industry by corporate giants, an intelligent federal procurement effort could drive up volume and drive down prices. This approach is particularly promising in the fields of photovoltaics and wind power.

A major barrier is the problem of establishing a high degree of public confidence in solar equipment. In several highly publicized instances, solar heating devices performed poorly. Usually the problem lay with faulty installation procedures, but occasionally the equipment itself has been the culprit. It is dangerous to demand too specific a set of standards and criteria for equipment in a young, dynamic industry, as this would tend to stifle cost-cutting innovations. But thorough training programs for solar technicians should be established, strict standards for installation should be enforced, and follow-up inspections should assure that the equipment is performing at full capacity. Suggestions for promoting this end include a federal warrantee program, the establishment of an industry liability pool, or a system in which the purchaser withholds the final installment of his payments pending one year of satisfactory operation. In addition to such programs, it is essential that public solar consumer advocates be available to warn people of possible problems and to advise them regarding promising remedies.

The farm is in a unique position to lead the United States into the solar era. It tends to be rich in sunlight, wind, and organic matter. Farmers tend to be skilled at maintaining equipment and willing to make capital investments that will save them money in the long run. The Agricultural Extension Service could be geared up to disseminate information on renewable energy resources. And the type of work done on the farm—heating buildings, drying grain, pumping water, making fertilizer—lends itself to renewable energy technologies. One of the most exciting solar prospects might be a comprehensive effort—modeled upon the rural electrification program—to bring solar power to the farm.

In some instances, existing energy enterprises—notably gas and electric utilities—have discriminated against solar consumers. For example, solar owners and business have been charged higher rates than they would have paid if they had no solar equipment—despite the fact that their heat storage capacity allows them to limit their consumption to relatively plentiful, inexpensive off-peak power. Such rate discrimination should be outlawed. Moreover, where dispersed solar equipment produces surplus electricity, the utility should in most cases be required to purchase this power at the full marginal cost of new power. Over time it is plausible that electric utilities will phase out of the generation business and become mostly vehicles for storage, transmission, and distribution.

Sunlight travels 93 million uninterrupted miles to reach the earth, but in the final couple of hundred feet it sometimes encounters trees or buildings. U.S. homeowners have a right to any sunlight that shines down vertically onto their property, but none does. The sun is always to the south of the U.S., and sunlight thus always slants over the property to the south. Since equipment to harness sunlight performs poorly in the shade, prospective solar purchasers should assure themselves of full access to the sun. Several innovative approaches to resolving this problem at the state and local level have been suggested.

Conclusion

The best energy investments available to the United States are in improved energy efficiency. Dollar for dollar, $250 billion dollars will save more energy if invested in conservation than it will produce if invested in *any* new sources of supply. However, no matter how much we conserve, we must develop new sources. Our entire society rests upon a diminishing fuel supply, and a transition must be made to more bountiful sources.

The choice of future sources will be largely determined by politics. Energy policy has tended to be established by making a choice and then arranging the necessary public subsidies and incentives to make the choice appear rational. About $150 billion in direct federal subsidies has been awarded to conventional energy sources over the last few decades. Far less would suffice to bring about the solar contributions outlined in this chapter.

We have the option today of making a genuine choice of what energy future we will bequeath our children. We can choose an assortment of fuels that are dangerous, vulnerable to disruption, and environmentally unsupportable. Or we can instead choose to turn increasingly toward the sun.

John H. Gibbons

10

Long-term Research Opportunities

Introduction

The time-focus of this chapter is toward the long-term future, beyond 1990, a time sufficiently distant that energy demand can be significantly influenced by innovations from scientific research and institutional change not available to us in contemporary society. Because this time horizon is so distant, we can only dimly perceive conditions that might obtain.

One common lesson from past history is that the future has always held surprises, good and bad, even for the most sage of us. Despite the inherent fuzziness of the future, there are a number of highly plausible assumptions that can be made and in doing so can guide our thinking about a relevant research and development agenda for the long term.

1. The price of energy, compared to other goods and services, will be substantially higher than at present. However, except for relatively short-term and emergency situations, real energy price will probably never be more than about four times higher than present price, because at that price level, given time, large quantities of renewable and long-term sustainable energy resources are economic. In the long term, we will not run out of energy—but we are rapidly running out of cheap energy. It is not unlikely that gas and oil prices (in real terms) will be double present prices by 2000.

JOHN H. GIBBONS *is director of the Environment Center and professor of physics at the University of Tennessee. In 1973–74 he was director of the Office of Energy Conservation, Federal Energy Administration. An editor of* Energy Systems and Policy, *Dr. Gibbons has written many works in the energy field.*

2. The dominant forms of energy will be the same as today: electricity, gas, and liquids (albeit derived from a variety of sources, increasingly nonfossil in origin).
3. As energy prices rise it will pay to alter patterns and efficiencies of energy use, substituting technological ingenuity and institutional innovation for energy resource consumption.

In the following sections we briefly examine the opportunities through research to provide conservation options which can be made available in future years to respond to higher energy costs. One important feature of this kind of research is that it should be responsive not so much to saving energy, per se, but rather to the broader goal of increasing overall *productivity* in our economy and our ability to compete in the world market, while at the same time remaining responsive to worldwide development needs, especially those of the developing nations. In this context we should *not* refer solely to labor productivity. Long ago in the U.S., manual labor was mostly replaced by energy and capital. A moment's consideration will dispel the notion that energy prices might rise so far as to cause men and women wielding picks and shovels to be economically competitive with bulldozers! At the same time one can easily defend the notion of technicians and engineers specializing in efficient use of energy replacing capital and labor in the energy production industry.

In all cases we consider energy only as a means to the end of desired goods and services in our economy. Energy, per se, is not a goal any more than other factors needed in our economy. Fortunately virtually *every* factor, including energy, used to produce goods and services (other than human ingenuity) is substitutable. It is the long-run flexibility and extent of this substitutability that is a central concern in this chapter.

Why Should There Be Tax-Supported Energy Conservation Research?

In the normal course of events in our competitive market society the needs as perceived by consumers are responded to by business in its constant search for profit. This system works best when market signals are clear, that is, when the full cost of a good or service is visible in its price. The energy market is obviously influenced not only by technology and energy resource availability, but also by a hierarchy of regulatory practices as well as international cartels.

Uncertainties in all of these areas make it extremely difficult to decide today about the economic viability of a new energy source or the optimum efficiency of a new energy consuming device or process (refrigerator, residence, industrial plant). Even if he is convinced that higher energy prices are coming, a businessman is hesitant to invest today in developing highly energy efficient devices that would only become competitive at considerably elevated energy prices. Yet even at today's energy prices we know that it is in the consumer's and even more the nation's best interest to move rapidly toward more efficient energy use (especially petroleum).

Since national self-interest in high energy productivity is great but private sector rewards are somewhat obscure at this time, and since time lags inherent between research and significant commercialization of research results are frequently long, we clearly have a situation where public investment in research and development (R&D) is justified. Public benefits of R&D in energy productivity include the following:

1. Energy savings can lighten the heavy and growing burden of balance of payments.
2. To the extent that long-term demand growth can be held down, existing supplies of energy resources will last that much longer, providing more time for developing their successors.
3. While energy conservation generally has beneficial environmental impacts due to fuel consumption avoided, there are nonetheless potentially important negative impacts associated with the conservation actions. For example, carbon fibers are used to produce strong, lightweight materials and, therefore, save energy. However, when released to the environment through wear and disposal, they can cause considerable damage, especially to electrical systems. (*Time Magazine*, March 13, 1978.) Improved understanding of such impacts and reduction of external costs is an important societal need.
4. New and improved energy efficient technologies should find important new markets in world trade, especially because most other nations have fewer energy resources than the U.S. and thereby face higher future energy costs.
5. A lower demand growth will enable a given amount of strategic oil reserve to last longer, increasing our resilience.
6. In the international context of depletion of fossil resources, it is the industrialized nations that are best capable of improving energy efficiency and holding down rate of energy price increase, hopefully enabling less developed nations to have a better chance to purchase fossil fuels at a low price and, therefore, be less dependent upon nuclear supplies for this development.

What Limits Energy Productivity in the Long Term?

NATURAL LAW

While "energy" is always conserved (first law of thermodynamics), it becomes less useful after each time we use it (second law). Thus the gambler's definition: the first law says that you can never win; the second law says that you can't even break even. For example, when natural gas is used to heat water *no* calories are lost, but the inherently very useful calories stored in the gas are converted to hot water and waste heat. In their new form they are far less "available" to do work. Therefore while energy, per se, is never lost, its use always results in degrading its *availability* to do work. That is one reason for the statement "there's no free lunch." As energy gets more expensive, the technologist designs machines that cause smaller amounts of high availability energy to be used for a given service rendered. The technologist also tries to match the quality of the energy source to the quality of energy needed to perform a service so that minimum availability is lost.

Another natural limit to energy efficiency relates the maximum theoretical efficiency of any cyclical process (e.g., a heat engine such as a steam boiler) to the combustion temperature (T_H) and the coolant (discharge) temperature (T_C). Well over a century ago a French engineer (Carnot) showed theoretically that the *maximum* efficiency (E) is: $E = 1 - (T_C/T_H)$ where temperature is measured in the Kelvin scale (32° F = O° C = 273° K). Clearly, unless the combustion temperature is extremely high and/or the discharge temperature extremely low, efficiencies are relatively small. It is possible using fossil fuels to generate combustion temperatures that result in very high efficiencies (fossil fuels, with air combustion, can burn as hot as 3000° F) but at such high temperatures and (steam) pressures that there are no materials that can readily be used to contain the steam above about 1300° F. Such limitations in power plant efficiency can be overcome by using a combined cycle wherein a "topping" cycle consisting, for example, of a gas turbine followed by a steam cycle is used. Even by combining such methods one is still practically constrained to Carnot maximum theoretical efficiencies of less than 70 percent, with 30 to 40 percent actually attained for modern electricity generation plants. However, as fuel prices rise and as technology improves we can expect to see "combined cycle" conversion with an overall Carnot efficiency of perhaps 60 percent. (See Chapter 7 for a discussion of future possibilities of higher efficiency in electricity generation.)

Heat pumps, air conditioners, refrigerators, and freezers are subject to the same laws of thermodynamics. It can be shown that while most heat pumps and air conditioners provide about twice as much heating or cooling energy as input energy (by extracting heat from outside air, or discharging heat into the outside air), they could *theoretically* supply about twenty times as much (hence they have an actual efficiency of only about 10 percent of Carnot efficiency). Equipment could be made with considerably improved energy efficiency and still be much lower than Carnot's limit; but each improvement in efficiency requires more investment (e.g., a more efficient heat radiator) and is, therefore, more expensive.

Since a large fraction of uses of energy actually only requires relatively low temperatures, it is clear that thermodynamics *matching* between energy sources and energy uses holds much promise. In other words, much better thermodynamic utilization can be obtained through a variety of techniques, including total energy systems, cogeneration, and low temperature storage. This notion is discussed in more detail below.

A Carnot-ary bypass? It is conceivable that some of the same functions (goods and services) now provided by a heat engine can be provided by some other mechanism. If so, the Carnot efficiency limits could be bypassed. For example, an electric motor powered by a chemical battery might be used to replace an internal combustion engine to power a car. These electrochemical reactions are not limited by Carnot's heat engine laws and can operate with very high efficiency. However, in thinking about such substitutes we must remember that other constraints will operate. For example, the electricity to recharge the battery must be generated somewhere, likely in a Carnot-limited device. Of course, it is possible that electricity could be generated from fuels via a hydrogen-oxygen fuel cell and again avoid the limitations of efficiency imposed on a thermal process. But where do the hydrogen and oxygen come from?

Other limitations to efficiency of energy consuming devices that are imposed by natural law include properties of materials such as heat conductivity, friction, and high temperature strength. For example:

1. a large fraction of energy required to power a car or truck above sixty miles per hour is due to wind resistance and rolling friction between the tires and the road;
2. the efficiency of a refrigerator is strongly dependent upon the properties of the heat exchanger or radiator which transfers heat into the outside air;
3. the efficiency of high temperature, high pressure combustion devices is limited by the loss of mechanical strength of boiler materials at high temperature; and
4. the finite resistance to heat transfer through insulating materials limits the conservation design in appliances, homes, and industrial processes where we wish to hold heat in—or out.

TECHNOLOGY

Clearly one of the most fundamental challenges to improving energy productivity lies in improvement of various properties of existing materials and processes and development of new materials and processes with desired special characteristics. As scientific understanding of the properties of materials grows, we have already witnessed profound impacts on technology and, in turn, on the amount of energy required to provide a given amenity. For example:

1. the progression from vacuum tubes to transistors and thence to large-scale integrated circuits has enabled the energy required for many functions in communications and computers to be reduced more than a millionfold in two decades;
2. the efficiency of electricity conversion into lighting has increased from less than 5 percent (incandescent) to more than 30 percent (alkali-halide) since 1950;
3. the energy required to produce a pound of aluminum from alumina has fallen from more than twelve kilowatt-hours per pound in 1950 to about half of that value now, with promise of considerably further improvement likely by using the new molten chloride process;
4. the energy required to produce low density polyethylene has been halved due to a new process;
5. research on producing a fire-safe diesel fuel resulted in a fuel that burns with 10 percent better economy and 50 percent less exhaust smoke; and
6. basic research on laser energy deposition in solids has resulted in a new, highly promising annealing process for making high grade semiconductor devices without need for an annealing furnace.

In all of these instances the notion of saving energy has been an important but not necessarily dominant one in spurring development of the processes. Indeed, many of these developments occurred while energy prices were falling. More important in inducing the developments has been the drive to improve overall quality, cost, and reliability of product. Another very significant factor has been the desire to improve the yield of final product compared to the consumption of raw material. Major advances in energy efficiency have occurred over the past several decades but some (e.g., conversion of coal-fired steam locomotives to diesel electric drive) were achieved at the cost of switching to less plentiful fuels. Nevertheless, close consideration of virtually every major energy consuming activity reveals major opportunities for energy to be replaced by technological advances. Of course the trade off in this replacement is (usually) increased capital cost.

COMPLEXITY AND INTERDEPENDENCE

As nations industrialize and urbanize, their citizens seem to accept conditions of increasing socioeconomic complexity in exchange for material benefits. The U.S. economy is now remarkably complex and interdependent. The U.S. consumer generally lives in an environment sustained by advanced technology. Perhaps the most important reason for this situation is the fact that increasing scale of operations (e.g., farming, communications, power production, transportation) tends to decrease unit production costs. Of course, at the same time this trend tends to have undesired but "external" side effects such as requiring workers to commute farther (at their own expense). In other words, increased interdependence between economic units can result in larger economic savings, albeit with lessened degrees of freedom.

In the case of energy, the "waste" energy from one user (e.g., a power plant) can be used as the "input" energy for another (e.g., heat for an apartment building). Similarly the purchased energy required to provide space conditioning in a house can be considerably decreased by installing a complex automatic electronic sensing and control system; or, transportation energy for urban commuter traffic can be drastically lowered by extensive use of carpools and vanpools as Robert Hemphill points out in Chapter 4. There are three conditions that tend to govern the extent to which these kinds of conservation tactics are attractive.

Energy price—As energy gets relatively more expensive, complexity and interdependence which enable more efficient energy use become more tolerable..

Reliability—We tend to accept complexity and interdependence more readily when the system can be counted on to operate without unwanted interruption. An apartment owner is not interested in being heated by "waste heat" from a power plant that cannot be counted on to operate steadily throughout the heating season. Improvement in reliability can be expected to continue to result from investments in science and technology; however, consumers are increasingly suspicious of complexities and new high technology; so there are important social discounts operating in this area that are certainly not well understood.

Proximity and Scale—Because of the past history of energy price, economies of scale in industry, and land use patterns of residential and industrial development, there are not many major opportunities for energy savings through increased interdependence in our *existing* system. Present sources of low-grade and waste heat, sufficient to heat and cool buildings, are generally sited too far from existing buildings to justify construction and transport costs of the heat. The extremely large size of power plants causes so much waste heat to be generated that markets within plausible range are oversaturated. As new capital investments are made, much improvement can be expected from energy conscious siting and sizing of energy production and other industrial plants which can result in considerably expanded opportunities for high efficiency "cascading" of energy use.

CONVENIENCE

Citizens, be they individual or corporate, value freedom. A petrochemical plant manager might save fuel bills by accepting waste heat from a power plant, but his hidden costs include threat of regulation, unscheduled interruptions, and inflexibility in future negotiations for fuel purchases. An urban commuter can save money in joining a vanpool, but in doing so he must abandon the freedom of schedule that would otherwise be available in driving his own car. An auto driven at fifty-five miles per hour might get 20 percent better mileage than one driven at seventy, but the occupant who highly values his time may choose the faster speed despite the losses. Thus "convenience" can be a strong but highly variable factor in limiting energy productivity.

ECONOMICS

Consumers seek to maximize their welfare by spending their income in ways that provide desired goods and services. Producers, in turn, seek ways to provide those goods and services in the most attractive (usually least cost) fashion in order to win over their competitors. As real energy prices rise, both producers and consumers attempt to readjust. Over a period of time comparable to capital stock turn-over, the energy required to provide a given amenity can be altered considerably. Because capital investments last a decade or more the producer/consumer must make some assumption about energy prices over the (future) lifetime of his investment—an uncertain target at best. It would probably be unwise to presume, as we have in the past, that future energy prices will be *lower* than present. Similarly it might be just as unwise to presume that energy prices are going to be an order of magnitude *higher* than at present. The challenge to R&D is both to lower the uncertainty about future energy costs and to provide options to allow much higher energy efficiency of use for minimum increases in capital cost, thereby desensitizing the investment from excursions of energy price.

Even during periods of declining real energy price (e.g., 1950–1970) the amount of energy required to provide a given good or service has decreased. Now that energy price trends are level-to-upward we would expect this past trend toward higher efficiency to continue, if not accelerate. Indeed, if past history has anything to teach in this regard, we would expect the energy price increase which has occurred since the embargo to induce considerable basic innovations that do not simply improve energy management but even supplant existing ways of providing amenities.

At the present time the energy market price and the price of energy intensive activities such as transportation are so distorted by public policies that it is difficult to discern real total costs. Therefore it is difficult to determine—much less project—the real cost of energy which can be used in trade off economic studies. Numerous analyses indicate that the investment required to *save* a unit of energy (e.g., insulation, improved process controls, high performance automobiles) while still providing the amenity is considerably less than the investment required to *add* a unit of energy (e.g., power plants, coal gasification plants, solar collectors). There is every reason to believe that further R&D in energy efficiency will continue to create additional cost-effective options.

Critique of Current R&D Programs in Conservation

After rising steadily for more than two decades, national expenditures of R&D leveled out dramatically in the late 1960s and remained virtually unchanged (in constant dollars) until 1975. Since then there has been a slight increase. Currently the number of scientists and engineers engaged in R&D per 10,000 population is about the same in the U.S. and Japan; the number is lower in Europe and Canada but considerably *higher* in the USSR.

Federal R&D focuses two-thirds of its efforts on defense and space-related studies, whereas Japan commits less than 10 percent of its efforts in these sectors. U.S. industry commitments to basic research have dropped since 1967 and in 1976 had dropped (constant dollars) to the level of commitments made in 1960. The U.S. reinvestment of new capital into R&D is low, whether measured by history or international comparison.

In the energy sector, both government and industry R&D investments have risen sharply since 1975. Government R&D in energy conservation was essentially nil prior to 1975 and nearly so in industry because energy prices had been declining for decades. Despite its recent high *rate* of growth, research in energy productivity is still not significant compared with commitments to energy supply. For example, the Fiscal Year 1979 Department of Energy budget for "conservation" amounts to about 8 percent of its total budget, up from 6.3 percent in FY78 (a growth of over 25 percent). These numbers can be compared with DOE's budget commitment to expand energy supplies which amounted (FY79) to 27 percent. It is important to understand that DOE's "conservation" budget includes electric utility system reliability, electricity transmission and distribution technology, power plant efficiency improvements, and other research areas traditionally thought of as supply-related. DOE's conservation budget can be disaggregated to three major categories.

Implementation of Action Programs—This nonresearch category includes weatherization for low-income families, assistance to states, audits for schools and hospitals, and consumer education. About two-thirds of DOE's conservation budget is spent in this area and most of these programs were instigated by Congressional actions. These programs are described in some detail by Robert Reisner in Chapter 11.

Technology Research, Development, and Demonstration—About one-third of the conservation budget is spent in this area. It is difficult to quantify the partition of these funds into "near-term" vis-à-vis "long-term" (post-1990) projects, but a rough division of four to one is apparent. The long-term work mostly consists of automotive (propulsion); electric utility (efficiency); community energy utilization (cascaded use); energy storage.

Nonhardware Research—A minuscule fraction of DOE's conservation budget is committed to this area (estimated liberally at 2 to 4 percent) of the conservation budget, or about 0.3 percent of the DOE budget. This category includes all nonhardware studies relevant to both near-term and far-term energy demand and conservation. It is arguable that hardware research is inherently more expensive than nonhardware (e.g., economic, institutional, psychological, marketing, demographic analyses) when measured in terms of cost per researcher, but it is not at all clear that the productivity (useful output per unit input) is any better—or even as good.

In its efforts to accelerate a near-term national shift to higher efficiency, DOE has placed primary conservation emphasis on research and programs that emphasize technology with high chance for payoff in terms of near-term savings of energy and money (e.g., before 1985). In this regard DOE has acted somewhat like an industrial entrepreneur and has chosen the lowest risk, highest rate-of-return actions. While such a federal strategy may be defended for the short run, it is inappropriate for the long run (the subject of this chapter) because a competitive market system should be able to attract private venture capital when potential returns are sufficiently high (or risks sufficiently low). This situation of R&D focus on the near term forms an interesting contrast to energy supply-research which tends to emphasize development of very long-term options such as fusion, breeders, and advanced solar electric concepts.

Another de facto constraint is worthy of mention: the tendency of Congress, OMB, and others to choose hardware over "soft" R&D. This tendency is reinforced by benefit/cost analytical approaches which tend to discount benefits of nonhardware actions (e.g., education and information to raise energy awareness of the public) simply because attendant savings are more difficult to quantify than those due to a new gadget.

Because of the nature of bureaucracies and their reward systems, it is felt to be so imperative that first demonstration projects "succeed" that over-conservatism results. By adopting private sector strategies into the federal conservation R&D program (already constrained by insufficient benefit/cost constraints and by the propensity of funders to choose highly visible hardware-type projects) several unfortunate consequences are nearly inescapable.

First, the criterion of near-term implementability tends to push R&D

funding into areas where the private sector should already have considerable self-interest. Unless very carefully orchestrated, such federal R&D activity could tend to retard rather than accelerate action in the private sector because of loss of market incentives such as proprietary know-how and patent protection.

Second, the criterion of high plausible benefit-cost again tends to put federal R&D funding in the same areas that one hopes would attract private investments. If a new conservation technology idea promises a likely 30 percent or greater rate-of-return on investment at present energy prices, then probably the best thing the government can do is to refuse to invest in that idea (except through its procurement processes), tempting though it may be, and instead provide a climate to encourage private sector investment and innovation.

Third, by emphasizing near-term development and high rate-of-return criteria the R&D programs naturally tend to concentrate on projects that are not speculative and that can be preplanned for major milestones. They also tend to be specific energy saving applications rather than generic approaches to energy productivity and resource stewardship. As a consequence, some of the most important new ideas could be ignored.

Fourth, the federal conservation R&D programs tend to deal exclusively, as do the supply R&D programs, with projects directly and specifically associated with energy. History teaches us that such a focus can well result in higher efficiency, such as a better insulated furnace, but that other major options with major energy implications might be missed (e.g., a new process that avoids the furnace entirely). In other words, crucial new options could naturally result from emphasis on increasing the *productivity* of all factors of production, including resources, rather than energy, per se.

Thus, for the longer term and in the national interest, it seems that public capital (R&D) should give priority to those situations in which the risks are so great but potential rewards are also great, the time for commercialization is so long, or potential returns based on *present* energy price are so low that private investment alone cannot rationally be expected to suffice. This kind of federal R&D can complement private sector R&D; however, because of the particular characteristics of such criteria, one must be careful not to let federal funding flow to the "turkeys." This simply means that the administration of this federal R&D must be sophisticated and, in general, be innovative and future-focused. The ultimate success of advances in knowledge, especially new technology, heavily depends upon implementation in the private sector. There is a long and complex distance between scientific feasibility and market implementability. It is therefore important to pay close attention to the entire process, beginning at the earliest possible stage (usually with the formulation or early monitoring of the research agenda).

The federal program should include not only new technology development and fundamental advances in science, but also a number of important frontier areas in the social and behavioral sciences, law, and administration. The fact is that energy is used by *people*—and we actually know very little about the basics of how our society operates, how it changes over time, and the opportunities for improved energy productivity through innovations in our social institutions.

A Federal R&D Agenda in Conservation

We know that limits imposed by natural law on the minimum amount of energy to provide the goods and services we enjoy are far, far below current practice. Typically the thermodynamic efficiency of energy use is less than a few percent, as compared with a few tens of percent for energy intensive production processes, such as that for producing electric power from fossil or nuclear energy. Clearly the *theoretical* opportunity for conservation through higher efficiency is dramatically large. Of considerably greater significance, of course, are the *practical* limits for conservation. These are dictated by both economics and by societal values.

The marketplace—if it is operating competitively—enables us to obtain a given amenity at minimum total cost, thus maximizing individual welfare. If *total* costs can be identified and are part of that process, then both individual and national welfare are taken into account. Of course personal preference and societal values also influence the choices we make in our market basket of goods and services. In the light of these facts, and in consideration of the roles expected of private sector interests, the *goals* of a federal R&D program in conservation should be to:

1. Provide new options to further substitute technological innovation for purchased energy consumption, especially oil—with emphasis on those options that cannot be logically expected from the private sector;
2. Develop new and modify existing institutional mechanisms (e.g., regulatory) to facilitate rather than retard the development and adaptation of energy and other resource conserving procedures and technologies;
3. Insure that such new options are either benign from an environmental and health aspect or at least have no more undesirable impacts than the alternative of using more energy;
4. Slow demand growth in a way that can lead to a long-term sustainable energy system, while continuing to help provide desired amenities.

Conservation R&D *strategies* to achieve these goals should include the following ten items.

1. Develop an extensive series of basic research activities aimed at improved fundamental understanding of physical, chemical, and biological properties, especially those of inherent interest in conservation; provide funding with long-term continuity similar to programs earlier supported by AEC and ERDA in physical research related to nuclear energy. Such properties should include, for example: high electrical and thermal resistivity and conductivity; high and low heat capacity; high chemical and thermal energy density; high strength-to-weight; low friction; high strength at elevated temperatures. Sometimes key advances come from unlikely activities. For example, an optical system designed for high efficiency light collection in a radiation detector used for high energy physics measurements has found direct and highly promising application in focusing-type solar energy collectors.

2. Structure an R&D program that explicitly encourages unsolicited proposals, especially in areas that are aimed at *new* ways of doing things vis-à-vis ways to incrementally improve existing ways of doing things. For example, while industry could be expected to do R&D to improve insulation around a furnace that is used to make ceramics, it might be the federal role to support speculative research aimed at developing new, low-temperature techniques to produce ceramics; and while insulated furnaces and metal scrap recycle can reduce energy in a manufacturing process, much more might be saved by shifting entirely over to power-metallurgy techniques for forming pieces. Such proposals should be especially solicited from universities, national laboratories, and small entrepreneural firms. Criteria for judgment of proposals should emphasize the potential for improved *productivity* in addition to energy impacts. Productivity is a measure of the capital, labor, and resources required to provide a given amenity. The most important feature advances for productivity in general as well as for energy will likely emerge from such investments in fundamental process development. The basic issue in this strategy is to test the extent to which we can substitute ingenuity (through technology and institutional innovations) for resource consumption in providing for human wants. Recent advances in information processing for monitoring, analysis, and optimized control of energy consuming equipment give a glimpse of vast future opportunities.

3. Develop increased opportunities for energy transformation and conversion at point of use. Of course this option has limited applicability in a nuclear economy, which is only capable of conversion in large central plants. The tendency toward electrification using massive, single purpose central station plants forecloses most opportunities to make productive use of the "waste" heat, which presently amounts to about two-thirds of input energy. Even with very advanced technology, about half of the potentially valuable heat energy would be thrown away at a power

plant. Most power plants are so large and sited so remotely from cities and industry that there is little chance to use the reject heat. If energy conversion were instead done in smaller units located close to point of use then there exist many more opportunities for productive use of more of the total energy resource. There are three general approaches: greater emphasis on producing gases and liquids and lesser emphasis on large central station generation of electricity; decentralized production of electricity (e.g., smaller sized fossil-fueled plants, direct solar, fuel cells); and siting of moderate-to-large size power plants close enough (less than fifty miles) to industrialized and urbanized areas that "waste" hot water can be economically piped to various consumers.

There are relatively few unique needs for electricity, a rather expensive form of energy. A gas-fired heat pump can heat and cool a building and simultaneously provide hot water and cooking, leaving only minor residual electrical demands. Gas can be obtained from natural sources or synthesized from coal, biomass, and biological wastes. It is not clear, however, in the case of gas synthesized from coal that coal-SNG-heat pumps are more efficient than coal-electricity-heat pumps. These trade off issues are complex and merit considerable study. Heat pump centered Integrated Community Energy Systems (ICES) have considerable long-term promise both for commercial and residential communities. One central challenge in developing this kind of diverse and dispersed energy system is to hold capital costs within economic bounds. *Therefore a central technological challenge is to learn how to do things efficiently on a small scale—a considerable departure from our past focus on achieving economy by going to larger scale.*

4. Support long-term research to improve our understanding of the role and future of energy and its substitutes in our economy, in the sense of both direct and indirect energy uses (e.g., energy consumed in the construction of buildings). Long-term relationships between employment (rate and type), income, GNP, capital investment, and energy need to be better understood. How long can our national welfare (related somehow to GNP) continue to expand at an appreciably higher rate than our energy demand? What changes will have to occur to make this relationship continue? As real income expands, how will increased "disposable" income be used? As our population growth slows down and relatively more people are in their middle and senior years, how will these demographic patterns and their accompanying changes in life style preferences affect energy demand? What will be the likely impacts of changes in energy price and policy on different regions of the U.S. and on consumers with different income? What are the energy implications of trends in land use, waste management, and transportation policy? Recent research, including both econometric and engineering analysis strongly indicates that economic growth and energy growth, while strongly

interrelated in the near term, can be substantially decoupled in the long run. The timing and ultimate extent of this process merit considerably more research and analysis.

5. Price and nonprice factors in consumer choice are another area in which considerably greater understanding is desirable. While it is defensible to presume that over the long run consumers, given clear market signals about total cost, will make minimum total cost decisions, this presumption is not necessarily valid. To the extent it is not valid, the potential for cost-effective conservation is compromised. Recall the farmer who was approached by the county agent with new information about how to improve his farming operations: "No need to bother me with that information, son; I ain't farming half as good as I know how to now!" Improved knowledge is needed about the processes of transfer, adoption, modification, and rejection of technological innovation by ultimate users. Similarly it seems crucial to develop increased understanding of the nature of political conflict over technological change and ways to resolve such conflicts in a productive manner. Otherwise it is extremely difficult to make plausible long-term projections. The structure of energy prices (especially electricity prices)—by time of day and season as well as by consumer size—will induce long-term responses by consumers which will affect the structure of demand. One should expect such responses to include some behavioral changes, but mostly technological adaptation (e.g., to take advantage of time-dependent rates). Extensive tests, designed to evaluate long-run rather than short-term responses, are needed in order to properly design rate structures.

6. There is a compelling argument that energy price should reflect its full social cost and also that incremental or replacement cost should be visible in energy sales to consumers. This principle is easier to defend than it is to devise practical ways to implement it, as pointed out in Chapter 7 by Bauer and Hirshberg. Thus, considerable research is needed to assess the full-scale costs of all major energy resources and to devise equitable ways to introduce incremental cost signals into the marketplace.

7. Identify possible future plausible spurs to energy demand. Much of the growth of energy demand since 1950 can be attributed to the advent of new, energy intensive consumer goods and services coupled with rising incomes and falling energy prices. Since 1950 many energy intensive products have been introduced to consumers and many are already well on their way toward market saturation. These include home freezers, television, and air conditioning. Further increases in automobile ownership and travel are limited simply because we now have almost one car for every driver and spend, on the average, almost one hour per day in a car. Illumination, refrigeration, water heating, and space heat-

ing are now universal in the U.S., and these markets are now mostly of the replacement type.

Some observers caution that these facts should not lead us to assume a consequent slowdown in the growth of energy demand. For example, airplane travel is apparently very price-elastic. Rather, they argue that new, energy intensive items will continue to emerge that will keep energy (especially electricity) demand growth high. One such notion is "outdoor space conditioning" and enclosed cities (although the latter could actually be less energy intensive than the alternative of separately space conditioning all the individual buildings). To be sure, there could be such developments, albeit presumably strongly coupled to energy price behavior. Consequently, more thought needs to be given to identifying those unfulfilled desires of consumers that are inherently energy intensive and to factor this into planning models that project future demand. In a similar vein, evolving social values could lead some segments of our society to seek *less* energy intensive life styles.

8. Effects of regulations and standards on long-term conservation potential are also matters of broad concern. Many regulatory processes were established when the U.S. policy was de facto to encourage energy use. Other policies, designed to meet other problems (e.g., to maintain competitiveness in freight transport or to keep paper pulp companies from competing with electric utilities), had the indirect but nonetheless real effect of fostering inefficient use of energy. Tax policy which favors current expenditure over capital investment has important, unfortunate consequences. Much research is needed to systematically delineate the energy and other resource implications of existing public policies and to devise productive changes which can be undertaken. Such policies include real estate taxation, freight rules, direct and indirect transportation subsidies, industry cogeneration and interchange agreements with utilities, and corporate tax policy with regard to R&D as well as capital investment.

9. Vulnerability and national security are undoubtedly affected by long-term conservation actions. As the nation moves toward lower energy intensiveness, it could become increasingly difficult to suddenly tighten our energy belts, as we did in 1973–74. It seems silly, however, to encourage waste just to keep some easily shedable "energy fat" in our systems! A counter argument is that if we shift to a much more efficient energy system then the loss of a given amount of energy will have less of an effect because energy would be a less important factor in our economy and a given strategic reserve will last longer. Clarification of the interrelationship between lowered demand growth and national energy supply strategies is obviously needed; yet comparatively little attention is presently being paid to this issue. These issues need to be better under-

stood to test the adequacy of our reserves and emergency procedures as well as to gauge the urgency of new energy supply development.

10. Our international relations with other industrialized countries, oil exporting countries, and developing nations are strongly impacted by our energy conservation R&D policies and progress. The U.S. call on world supplies of energy will be impacted by our domestic developments; thus other oil importing nations watch our policies with great interest. The U.S. has endured a zero domestic energy supply growth since 1968 by turning to imports while at the same time doing little to slack its increasing energy thirst. Energy importing countries provide extremely important market opportunities for U.S. conservation technology. Similarly, U.S. efforts to assist lesser developed countries in their economic development could be profoundly benefited by energy efficient innovations that are adaptable by less technically sophisticated societies.

It would not be hard to prepare a lengthy list of specific research projects to illustrate the strategies we have outlined. Rather than attempt to do that I simply list below a few examples, mostly technology-oriented, of long-term conservation R&D opportunities.

1. Substitute information technology for energy consumption. The rapidly evolving technology of miniaturization of information processing and storage has numerous potential applications for decreasing energy demand for a given amenity. Examples include space conditioning, and combustion and process control.
2. Partially replace travel with communications. More and more of the work world, as well as leisure (e.g., television), is dominated by information transfer. If we could learn—in a sense both of technology and behavior—to substitute a significant portion of travel with electronic communication, the resulting direct impact on energy demand would be very great. The present rate of advancement of electronics and data communications is so great that it is difficult to say much more at this time because feats in this area that seem incredible today could well be commonplace within a decade.
3. Improve performance for ground and air transport. Long-term R&D should enable a 50 to 100 percent improvement in jet and automobile engine performance over those that are used today. Major gains should also be possible as a result of development of stronger but lighter weight materials.
4. Plausible advance in powder metallurgy could greatly reduce energy required for production of forged metal products—and, in addition, considerably reduce metal losses during processing.
5. Substitution of electric induction heating for direct fuel heating in certain industrial processes could cut energy losses since the heat can be applied where most desired, can lead to efficient variable load processing, and can

result (e.g., in glassmaking) in reduced pollution emissions as well as improved by-product recovery.

6. It should not only be possible but also cost-effective to construct buildings that are attractive and functional but require no more than one-fourth as much energy for heating, cooling, and lighting as is used today. These gains should be possible by a combination of energy conscious design; careful construction; advanced technology in heating, cooling, and ventilation, including interseasonal thermal storage; and sophisticated controls; and compromises in standards of performance and reliability.
7. While naturally focusing on developing new options, conservation research should also identify the long-run importance of *existing* options that, if ignored, might be abandoned prematurely; for example, existing rights-of-way that could be utilized for ground transport and district heating; or land parcels that could be dedicated to co-located power plants and industrial parks.
8. There are many long-term research opportunities associated with the electricity sector, especially in the areas of power plant conversion efficiency, cogeneration, and demand management at the load point.

Conclusions

Only a cursory examination of the way we use energy in contemporary society reveals major opportunities for improvement in efficiency, especially over the long term, corresponding to replacement of capital stock. A similar examination of limitations to such improvements that are imposed by natural law discloses that only in a few instances have we begun to approach these limits. Trade off studies between energy efficiency of use and energy price indicate that large improvements in efficiency of energy using devices are possible for relatively small increases in production cost. One role of research is to constantly roll back the limit to efficiency that is compatible with a given energy price. This is ultimately reflected in the amount of energy required to sustain a given GNP. The ratio of energy consumed per unit of industrial production has slowly declined since early in this century. Aggressive research should enable that trend to continue, if not accelerate.

Another role of long-term research is to provide improved insight into the nature and trend of our energy consuming society—the choices we make, how and why we adopt or reject technology. Similarly, research helps us to understand the social and economic implications of energy conserving actions in terms of such important issues as: employment, social equity, freedom of choice, resilience in the face of emergencies, international security, environment and health, and urgency for new supply development. Since about half of our population that will be living in 2010 is not born yet, it is likely that the social and cultural values of that year will be different from those we hold today.

While there are obvious private market incentives for research in conservation focused on near-term opportunities, many of the major long-term opportunities are beyond the reach of attracting private capital. At the same time, the national need for research in various conservation subjects, which for reasons of both time lag and insufficient price signals cannot be expected to be reflected in private market calculus, merits substantial and continuing public support. There are several well-reasoned projections and forecasts about the future, and they differ substantially! While R&D will likely shed little light on which future is more likely, it will help understand and shape the future by providing a broader array of technical options to choose from and also a clearer understanding of the future implications of present trends and decisions.

All of the strategies for conservation discussed in this chapter find both near-term and long-term application. Even though we have given serious thought to these matters since 1974, it is clear that the surface is scarcely scratched. Ultimately we may translate the specific issue of energy conservation goals into a more general set: *to what extent can mankind substitute ingenuity for resource consumption?* If so, our measure of progress may become less that of counting how *much* we consume and more that of how *little* we require to provide a given level of amenities.

Robert A. F. Reisner

11

The Federal Government's Supporting Role in State and Local Conservation Programs

Previous chapters have analyzed the impact of energy price trends and technological developments on energy use in the U.S. Various regulatory approaches have also been explored. This chapter will discuss a different dimension of energy conservation policy. Here the focus is on implementation. The premise and performance of federal programs for state and local recipients are reviewed. These grant programs are intended to encourage adoption of energy conservation techniques; their effectiveness is an important budget and policy issue.

In the years immediately following the embargo, the federal government has created a number of different state and local grant programs. These programs have grown surprisingly large to the point where, in the 1980s, they may receive as much as a half a billion dollars in funding annually. The objective of these programs is to build and support a network in states and communities to reach small energy consumers to stimulate energy conservation programs throughout the country. Congress found that the states represent a unique delivery mechanism to reach energy consumers with face to face conservation services. The problem is that state grant programs may be an especially expensive type of program. Energy conservation assistance in this form could grow to re-

---

ROBERT A. F. REISNER *is a Principal in the public policy consulting firm ICF, Inc. and a lecturer at Yale University's School of Organization and Management. He has been Associate Deputy Administrator of the Federal Energy Administration.*

quire a budget greater than all other conservation activities combined.

The trends in the growth of state grant programs suggest some very interesting questions. Is this growing reliance on grant programs an effective strategy to encourage energy conservation among small energy users? Will these grant programs achieve results that will change the pattern of energy consumption during the next decade? Should these programs be better organized to provide energy conservation services? What should be the role of federal, state, and local efforts? These questions and others are raised by even a cursory review of the new conservation grant programs. This chapter first reviews the federal grant programs; it considers their problems; it reviews state and local programs which are already operating; and it outlines alternative paths which might be followed in reorganizing the grant programs to make them more efficient.

A review of the programs and the issues shows three interesting things. First, the early growth of the state grant programs has helped stimulate a vast array of quite sizable and growing programs. Second, the next big job in state and local energy conservation will be to rationalize and consolidate the early investments. Third, the second step will not be easy.

This chapter will also show that the complicated nature of the state energy conservation programs is caused partly by the laws that have created them; partly by problems in the way in which they are run. These characteristics taken together threaten to make the federal effort ineffective.

But, even if there are problems in starting up the federal programs, this chapter will show that there is already a broad array of small-scale energy conservation activity going on in the states. If it is an important public policy to stimulate local action and to broaden the application of successful conservation programs in the states, difficult choices are posed for federal policy-makers. On the one hand, the grant programs should be reorganized to make them more efficient. On the other hand, evaluation that selects promising programs from among the diverse experiments currently under way is difficult. Tightening management while retaining experimental flexibility will be a hard balance to achieve. The near-term value of government program support to improving energy efficiency depends upon making such difficult choices.

The Federal Grant Programs—Their Growing Complexity

There are at least six separate existing and planned energy conservation grant programs. Table 1 illustrates the various programs and points up their growing size in comparison with the remainder of the conservation budget. These similar grant programs are complicated and their overlapping purposes are confusing. Nevertheless, important differences do exist which make a brief review important.

TABLE 1. BUDGET AUTHORIZATIONS AND APPROPRIATIONS (MILLIONS OF DOLLARS)

	Fiscal Years				
	1976	*1977*	*1978*	*1979*	*1980*
State Energy Conservation Program (EPCA)					
Authorized	50	50	50		
Appropriated	5	23	47.8		
Requested for FY 79				47.8	
Supplemental State Energy Conservation Program (EPCA)					
Authorized	—	25	40	40	
Appropriated	—	12	23.7		
Requested for FY 79				10	
DOE Weatherization Program					
Authorized	—	55	65	80	
Appropriated	—	27	64.1		
Requested for FY 79				199	
Additional Request (NEP)			65	120	
Energy Extension Service					
Authorized	—	—	7.5		
Appropriated	—	7.5	7.5		
Requested for FY 79				25	
Schools and Hospitals Program					
Authorized			200	300	400
Appropriated			200 [1]		
Requested for FY 79				300.5	
Local Government Buildings					
Authorized			25	40	
Appropriated			25 [1]		
Requested for FY 79				32.5 [2]	
CSA Weatherization Program					
Authorized [6]					
Appropriated	110 [3]		103 [4]		
Requested for FY 79				0 [5]	

Source: DOE, August 1978

[1] Appropriated but cannot spend until NEA is passed.

[2] Includes amendment for 22.5 M.

[3] Includes amendment for 82.5 M.

[1] Includes programing of 38 M of FY 77 money.

[5] Zero in the President's budget but Senator Kennedy introduced an amendment to DOE's Appropriations Bill to give CSA half (99.5 M) of the DOE money.

[6] Authorizations for CSA are open ended.

The State Energy Conservation Program (SECP) which was created by the first piece of comprehensive energy legislation, the Energy Policy and Conservation Act (EPCA), is the basic program of support for the state energy offices. A second program, the Supplemental State Energy Conservation Program (SSECP) was established three-quarters of a year later, in 1976, by the Energy Conservation and Production Act that extends the life of the Federal Energy Administration (FEA). The SECP contains five mandatory activities that a state must undertake to receive funds. The SSECP differs by focusing primarily on the techniques of publicity, intergovernmental coordination, and energy audits as a means for providing energy conservation services. While the first program established state energy offices and required that they undertake certain activities to help state governments save energy, the second program was intended to be a catchall piece of legislation to include proposals which had long foundered in Congress because of political opposition. So, for example, the extremely significant buildings standards requirements for states to maintain energy efficient building codes and the provision for FEA to establish a state grant weatherization program were included in the bill that created the SSECP.

The SECP and the SSECP were written by a Democratic Congress concerned that the Republican administration would not implement tough energy conservation programs. The Nixon and Ford administrations had often stated their opposition to programs not relying on market mechanisms. For budget reasons, the Office of Management and Budget (OMB) was also opposed to the development of such programs. The establishment of state government offices to promote regulation, OMB argued, would lead eventually to requirements for more staff and budget support and would encourage expensive, new regulations

The creation of two other major grant programs, the Weatherization Program and the Energy Extension Service (EES), were actions of a different kind. The SECP and the SSECP were intended to create and support state energy offices which would be responsive to the Department of Energy (DOE). The Weatherization Program of the Community Services Administration (CSA), however, is basically aimed at providing support for the community agencies of the CSA originally formed under the Office of Economic Opportunity. An entirely separate Weatherization Program in DOE was created by the ECPA legislation which established another state grant program. In the DOE program, funding is given to the state except in the event that a state's application is denied. Under these circumstances, a local government or community action agency of the CSA may apply for direct funding.

The Energy Extension Service was intended to support existing agencies having a capability to provide energy extension services. For example, in a given state the EES might fund the Cooperative Extension Service (CES) to provide energy workshops and home energy audits. In fact, CES, a Department of Agriculture (USDA) agency, may have been intended by some original sponsors of the concept of the EES as one of the primary recipients of Energy Extension Service funding nationally. Hence, both the Weatherization Program and the Energy Extension Service were created with the significant impetus of lobbying efforts from their political constituent agencies. In both cases, Congress gave primary responsibility to the Department of Energy, not to parent HEW (CSA) or USDA (CES) to establish the programs. Not surprisingly, both programs have seen a great deal of pushing and pulling from various bureaucracies in their short lives. Even within the Department of Energy, the various divisions responsible for the different conservation programs have from time to time contested the way in which they are being run.

The conservation grant programs may become even more complex. The National Energy Plan (NEP) proposed by President Carter in April of 1977 includes a major grant program to help insulate hospitals and schools and other public buildings. These public facilities are known to be among the most poorly insulated structures in the country. But in addition to reaching a key target of energy conservation, this new program will also reach a key political target. A new categorical grant program will be created through the hospitals and schools initiative. The new assistance program to help insulate schools will provide a means (albeit on a matching funds basis) for supporting local budgets and will assist municipalities in coping with the financial strains of higher energy costs.

The Project Conserve program is another energy conservation initiative that falls outside of the mainline programs. This program was initiated under DOE's general energy conservation legislative mandate. It was intended to provide computer home audit services as a means of assessing the energy costs of individual homes. Because of opposition from OMB to the costs and concepts of a direct service federal delivery program—where the federal government, through means of a computer, would interact individually with millions of homeowners—the Project Conserve computer home audit concept was initially implemented in only two states. In an additional fourteen states, the federal program was modified so that it could be implemented with a manual that would be sent to homeowners. Project Conserve's emphasis on home audits was picked up by the SSECP and the EES and weatherization programs. Though intended as a research effort, the Project Conserve concept is still alive and in some states is viewed as another federal program.

In sum, at least four (perhaps six) operative programs were created after the embargo. There are even more if all of the similar programs with an outreach capability are taken into account. They have established a confusing pattern of interactions in the states. But at the same time, they are politically popular. The experience of federal programs in other domestic policy fields is that categorical grant programs of this kind continue to be popular because they provide a means for communities to attain federal resources in order to support local projects. While a consolidated block grant program often may be easier for the state grant recipient to administer because of its comparative simplicity, such programs have less political appeal than the categorical programs over which individual congressmen can exercise greater control.

Hence, in Washington, consolidated grant programs are inevitably more difficult for Congress to pass than categorical ones. Even with the complex structure of the existing energy conservation program, there may still be pressure to create new categorical programs to fill existing gaps in service. The political popularity of new programs suggests that specialization may be a popular direction for programs to follow, whatever the possible implications for program management. Reform legislation has been under serious consideration in Congress, but consolidation will be difficult to achieve. Effectively implementing a reorganization into a consolidated, simplified, conservation grants approach will require exceptional political leadership.

The Grants Programs in Operation—the Need for Consolidation

To see the problem caused by duplicative programs more clearly, it is helpful to examine what is actually going on in the field where the grants are in operation. The array of similar federal programs providing assistance to states and communities has understandably been a source of confusion for nearly all concerned. For the recipient of federal grants, confusing signals are generated by the existence of competitive sources of support. Parallel programs naturally cause duplication and impose conflicting requirements on the states. For a state, the day-to-day reality of dealing with the federal government consists of regulations, guidelines, authorizations, appropriations, apportionments, monitoring reports, quarterly reports, visits from federal monitors, liaisons, evaluators, and so on.

These details of the implementation of a federal program have a critical impact on the way in which state and local government recipients of aid actually behave. The rules are often complicated even in dealing with one program. When, for example, conservation programs are managed by several parts of the federal government, the inevitable conclusion is a complex, often inconsistent set of rules. A state often finds that when federal guidance is being transmitted through several bureaucratic pipelines, each with their own constituency and corresponding agency in the state, communication between Washington and the field can become very complicated.

Examples of conflicting directions abound in the histories of the grant programs. One conflict took place when the EES pilot program and the SECP programs were both being implemented. The EES, a former Energy Research and Development Administration program, used commercial banks to disperse funds. But the SECP, a former Federal Energy Administration program, used Department of Treasury local dispersing offices. Though both programs were clearly to be DOE programs, the respective business offices insisted that their traditional dispersing method be used. They both won. Not surprisingly, confusion was created in the states that had both programs and two dispersing methods.

Another example is the effort to coordinate the operations of EES, SECP, and SSECP. Naturally, many states having the three programs sought to consolidate functions, such as advertising, in a central place. Different programs could have assumed responsibility for different functions. Though sensible in concept, problems developed because the SECP (and SSECP) programs were run through the DOE regional offices, while the EES pilot program was run from Washington. Approval procedures operated differently and frequently the advertising was not ready at the time that the service programs were ready to deliver services. Other examples of conflicts are numerous. Different funding cycles, reporting requirements, and philosophies cause conflict and duplication. Even if one program were created from six, there would very likely be management problems during the transition.

Reform can itself be disruptive. Even where there is consensus that a change is needed, as there has been in the case of the conservation programs, reorganization can be very threatening to program managers. Uncertainties encourage bureaucratic infighting over control. Consolidation often means that one program is subsumed within another. At best, consolidation proposals contribute to massive uncertainty for the staff. The argument over which program controls another usually revolves around code words that denote turf and organizational nuance. As a result, the federal programs engaged in bureaucratic fights are continuously communicating with one another, OMB, and Congress in a "bureaucratic" language which confuses their target audiences and clients.

Will the review and oversight process prevail and simplify the six or more grant programs? Or will the power of the growing constituency of grant recipients encourage further fragmentation and specialization? Political science does not offer any clear basis for prediction. In the past decade, phases of program proliferation have been followed by consolidation only to be followed again by new program development. One clear conclusion, however, is that the programs have been reformed only after extended debates and political compromise.

Other Government and Private Programs

The federally supported grants programs are, in fact, only a part of the conservation activity operating at the state and local level. The other new programs include tax incentives, private sector initiatives, and state and local independent efforts.

The tax incentives proposed by President Carter as part of the National Energy Plan were a renewed attempt to use the Tax Code to encourage energy consumers to invest in energy conservation measures. The significance of tax proposals may outweigh all other initiatives taken together. These incentives are not limited by budgetary considerations in the same way in which a grant program is. The Carter tax proposals for individuals were not only potentially very effective, they were popular. Congressional reaction to the tax proposal was immediately favorable. While other provisions of the Carter NEP were more controversial, the tax incentives were agreed upon within months of the President's submission of his National Energy Plan.

Energy users may also be reached by the widespread and growing private sector involvement in providing energy conservation services for a fee or as a business investment to develop customers. New firms specializing in audits of homes and businesses, vendors of insulation and conservation equipment, consulting engineers, and insulation contractors are all seeking to develop aspects of the conservation retrofit market. Additionally, institutions such as utilities and conventional lenders have developed new programs in recent years to encourage energy conservation at the state and local level. Literally dozens of utilities have developed programs to provide energy conservation advice to their consumers. In some cases, these audits have become quite popular with the overall result that many more audits have been conducted by utilities than by the government. Some public service commissions have introduced requirements that utilities within their states provide conservation assistance to their consumers as a matter of law. These requirements are often levied as a part of rate setting hearings in which utilities are granted rate increases with the condition that they provide new services in energy conservation to their consumers.

Utilities, in some instances, have also provided loans to their customers to assist them in making energy conservation investments in their homes. Under one popular approach, customers pay back the loan in their monthly bills. Problems of customer protection have inhibited full-scale adoption of this approach to encouraging energy conservation. How the customer is to be protected in selecting insulation contractors and the advisability of having the utilities become, in effect, banks making home improvement loans, has been a major stumbling block to complete acceptance of this kind of a program. In Chapter 8 of this book Denis

Hayes describes a similar approach for financing solar equipment which has met with considerable acceptance.

State Conservation Programs—Some Examples

Different states have taken varied approaches toward initiating energy programs. In Connecticut, for example, the state energy program was initiated by forceful state leadership several years ago. The state has been very successful in obtaining federal grants. Programs, including a Project Conserve Program, an Energy Extension Service Program, sizeable numbers of HUD solar demonstration sites, and CSA winter crisis support funds, all supplemented the basic weatherization, State Energy Conservation, and Supplemental State Energy Conservation Program. With strong state leadership and active public utility programs in some sections of the state, the Connecticut approach has been to run almost the entire state program through the state government. This approach has unfortunately contributed to a slow start-up because of the obstacles presented by a slow moving bureaucracy and civil service hiring problems. The experimental early design of the Connecticut program made hiring staff somewhat more difficult since there were initially few permanent jobs which fit into the civil service classification system. In Massachusetts, on the other hand, implementation proceeded more rapidly. In part, this was due to widespread public interest in energy conservation and highly active local communities. For this reason, conservation actions taken by the state, even with limited financial support from the federal government, out-paced the actions supported by the federal program. In programs which actually were run by the state, local community participation contributed to extensive interaction with program recipients. Channels for the distribution of materials, for example, are more active in Massachusetts because of the reliance of the community on universities and local organizations to extend the capabilities of the state.

California presents still another example of extremely active community support. In California, the size of the state government and the wealth of resources available within the state combined with aggressive community efforts to create state programs, almost as an alternative to federal action. The fundamental political climate in California in the years following the embargo cannot be ignored in assessing energy conservation actions. Local communities initiated energy programs on their own. Several communities throughout the state sought a less energy intensive life style, and the state government developed legislation to support energy conservation and solar investments.

Cities in California encouraged reinvestment in inner city communities. Local governments were exceptionally supportive of the changing life style of a more energy conscious culture. In part, the governor's support for the "small is beautiful" movement and the publicity given to examples of community action supporting this theme contributed to the movement. This fundamental political support combined with the basic inclination of the state government to support innovative programs provided an alternative to Washington. Thus, even without a proliferation of federal grants as in Connecticut, the state government played a major role in stimulating energy conservation programs. In the post-Jarvis-Gann era, it will be interesting to see whether the state continues to support such potentially counter-growth initiatives.

A producing state like Texas presents a different model from the examples of Connecticut, Massachusetts, and California. If the first state represents the model of strong leadership from a state goverment; and then the second and third show examples of state government with exceptionally active community sponsored energy conservation programs (the first in a state with constrained state resources, and the second in a state with, until recently, relatively abundant state resources); the producer states, of which Texas is perhaps the most striking example, have not been strong supporters of energy conservation programs. In fact, in Texas the universities have represented some of the strongest examples of initiating conservation programs outside of state government.

We can generalize from this brief review of alternative energy conservation approaches at the state level that community support for energy conservation initiatives has been an important factor in contributing to the early start-up of innovative activity associated with the grant programs. The presence of financial resources is also important. Thus, the early grants in states with resources and community support will likely show early results.

The presence of energy minded communities is an important factor often overlooked in the development of successful programs. Wisconsin and Minnesota are generally acknowledged to be among the most effectively run state governments in the country. Local communities in these states are supportive of conservation. The existence of high energy prices in combination with the community support of an energy minded and effective state government provides the basis for an active state energy conservation program. Significantly, these variables—energy mindedness, effectiveness of state government administration—the price of energy and the relative location of energy resources, are different in each state, contributing to differences among the performance of state governments in the aftermath of the energy embargo.

Local communities have also been laboratories for development of important innovations in measuring the effectiveness of conservation services. HUD, for example, supported significant evaluation experiments in Annandale, California, and in Portland, Oregon, in addition to support that the department provided to the state of Massachusetts. One of the important conclusions of this research, a finding that is matched throughout the country in other energy conservation programs, is that the capacity of state and local governments to analyze their energy use and to take action is directly related to their ability to devote staff time to program design and evaluation systems. For many city governments, one of the most important steps that can be taken to encourage investment in energy conservation is to simply establish an accounting system for energy use.

Judging Effectiveness: The Dilemma of Energy Savings

Unfortunately, the practice of determining which programs will lead to effective energy conservation has not kept pace with the rapid growth of conservation expenditures. Hence, there is no consensus that any of the federal programs represents an effective allocation of resources. This section considers the problems inherent in judging successful conservation grant programs. These problems include difficulties in measurement, attribution, verification, and in making comparisons. Finally, the implications of these problems for evaluation are considered in terms of the underlying issue of choosing the most effective organization for these grant programs in the future.

The first problem with judging success is that the programs have not demonstrated impressive accomplishments to date. The first few years of the state programs have been devoted largely to writing regulations, developing plans, and hiring. Programs for homeowners which provide services through hotlines and clearinghouses, audits, and workshops have only recently been initiated. Four years after the embargo, energy users were just being reached. The argument that it is too soon for anyone to know how much energy these programs may potentially save may be valid. An example helps illustrate the problem.

The concept of measuring energy conservation performance was an essential feature of the first energy grants program—the State Energy Conservation Program. It did not work very well. When the SECP was first designed, the grant formula was structured to provide assistance to the states in accordance with their energy savings. The states which were successful in implementing programs that produced savings were to be rewarded by receiving a bigger share of the pool of funds.

The Office of Management and Budget under the Ford Administration was strongly opposed to the kind of detailed interaction that such a procedure would entail. OMB's opposition was based on the fact that the energy savings were not accurately measurable. Hence, the formulas that would be used to establish the levels of the state grants would inevitably require detailed federal interaction with the state programs to assess their savings. OMB's opposition resulted in adoption of a compromise funding formula for the SECP program in which part of the state's funding could be based upon its accomplishments in saving energy and part on a formula involving the population and energy use within the state.

The adoption of a hybrid approach, balancing population, energy use, and savings lessened the emphasis of the state grant programs on determining where the greatest energy savings could be achieved. Greater emphasis was placed upon the delivery of services and the development of plans. But the need to measure energy savings has not been eliminated. States demonstrating energy savings will be permitted to share a bonus.

The experiences of the states in implementing the program have been less than optimal. Many state energy planners feel that the SECP's emphasis on quantitative measures has created a kind of "body count" mentality among the states who are competing for funds. They are forced by the competition to show savings that range beyond verifiable accomplishment. As a result, the numbers are widely viewed as meaningless. Moreover, there is no incentive to report savings accurately—just the reverse. Some states have even found that because of internal DOE data problems, they could qualify for incentive funding on the basis of imaginary savings.

The reasons why it is so difficult to measure success by quantifying energy savings are not difficult to see. One problem is that many of the factors which contribute to energy conservation actions lie beyond a program's control—the level of energy prices and the availability of energy technology, for example.

If energy prices are permitted to rise and technology for energy conservation becomes more widely available to individual homeowners, the demand for energy conservation services is bound to increase significantly. These factors, however, are beyond the control of individual states. Another problem with evaluation involves verification. Where a program may have demonstrated that it has satisfied significant consumer demand (for information, for example), it is still hard to show that it has achieved significant energy savings. Did the consumers take appropriate action? How can federal officials be certain? What baseline assumptions were made about the potential consumption pattern with no conservation program? Where there have been investments in insulation, have they actually achieved the promised savings? End use data provide few answers in the short term. The programs are too limited to be measured in gross terms and too submerged among other forces to be able to demonstrate results except through self-reporting.

These questions of savings measurement are highly charged politically. Evaluation questions are being asked just at a time when the funding of these programs is becoming more controversial. As programs have grown so have "special interests." As each funding cycle is passed, the states and other beneficiaries of the various conservation programs will increasingly represent a major lobby group. Charges that these programs are not effective are already met with strong protests. Measurement of success is both uncertain and sensitive politically.

Even if measurement were dependable, there is still very little basis for making comparisons. Programs differ widely among states. The diversity of the country itself is a fundamental factor explaining the variety of approaches that have been used. The 1973 embargo affected some states deeply and cut gasoline supplies. In other states there were surpluses and thus the shortage had no effect on energy consciousness. In the case of Texas and Arizona, for example, the first state was barely touched by the embargo and the second was practically crippled. The existence of supply lines, the configuration of retail oil companies, and the dependence of local areas on various fuel sources are such that there were widely different effects in the crisis. When two states can have such completely different experiences in a shortage, it is not surprising that there would be widely varying levels of energy consciousness throughout the country.

Despite the measurement problem, there is a broad consensus in favor of consolidating the grant programs and rationalizing competing programs. The consolidation will most likely emphasize making the programs more understandable. Change will require wide participation including that of the Congress since many of the energy conservation programs have been established by statute. In the consolidation process, some extremely significant decisions must inevitably be made, such as which programs and early initiatives will be retained and which will be eliminated. Careers and reputations will be at stake. The choices will not be easy.

The general case for consolidation is not difficult to make. But, the process of implementing a simplified structure will be hard even if a blueprint were in existence. For example, how will the various roles of the Community Services Administration and the Agricultural Extension Service be reconciled with simpler and more consolidated state grant programs? A simpler program might rely upon a central state energy office. But such an approach would inevitably tend to favor the new state level energy bureaucracies rather than other organizations which may be more effective in some states. Even if there were advantages, which could be measured in energy savings, to using the capabilities of existing organizations in the field, they may be precluded from playing the central role in administering conservation programs in a structure that seeks a standard national pattern. Should the states be permitted to have two organizations which are both responsible for major elements of conservation programs? How much competition is beneficial at the state level? When do infighting and debate detract from delivering services?

These organizational choices are made more difficult because there are a number of diverse goals contained in the state energy programs. The goals include providing services for low-income populations, addressing the needs of small consumers, encouraging the involvement of existing organizations, etc., which might not have high priority in a streamlined state energy conservation program.

Important trade offs will have to be made among the various objectives of the existing programs if they are to be successfully integrated. Choosing lead programs will favor some energy conservation strategies over others. Saving energy may sometimes best be accomplished by making investments in large sources of energy use such as factories or large users of energy for transportation. But addressing these energy uses may lead to energy expenditures which are more beneficial to commercial or public sector institutions than to individuals. The goal of mitigating the impact of high energy prices on small consumers could be an early casualty of the reform of the admittedly jumbled set of existing programs. Innovative programs may also be cut if the key goal is demonstrating savings.

Obviously, there must be some balance among these variations. Competing programs should not all be receiving federal support. But innovation should not be eliminated. The means for selecting the appropriate balance among the alternative objectives of the existing federal programs is an intensely political process. The most difficult issue is whether decisions should be made upon explicit grounds of judging conservation savings or whether other softer criteria should have equal importance.

The policy objective, despite the measurement problem, should be to encourage programs that save energy. Should the nation depend upon state grant programs as a delivery mechanism for achieving energy conservation if they cannot demonstrate energy savings? Herein lies the dilemma. The practitioners who are responsible for running these programs, of course, believe that they have intrinsic merit. Yet there is great reluctance on the part of even the strongest advocates to assert that grant programs can achieve measurable energy savings in the short term and should be judged primarily on that criterion.

The Policy Implications

The problems in evaluating the conservation grant programs point up a number of important considerations for policy-makers who must decide how much money to spend on grants and how soon. First, regardless of the difficulty of the decision, the size of the grant programs will make it imperative that judgments be made. The future budget for the programs discussed here could equal a sum greater than that budgeted for all other energy activities in the decade of the eighties. Second, the organization of these programs has been exceptionally complicated.

From the viewpoint of the recipients of services, the federal structure is confusing. The parallel organizations which have been responsible for the energy programs encourage duplication and competition.

If the energy conservation grant programs are shown to be expensive and inefficient, they will not be judged favorably. From the perspective of achieving the maximum energy savings of getting services delivered to consumers who need assistance and from the perspective of communicating clearly with the states, the grant programs may fail the tests of cost benefit analysis that are usually applied to major expenditures of federal resources. In the next few years, budget analysts who review these programs are bound to ask what is being achieved with the money? Program managers who realize that resources directed into these state grant programs might otherwise come to them may not even hold their criticism that long.

But early criticism of the management process may have been directed to short-term phenomena. In the early years of the development of the state grant programs, a comparison of costs and achievements will naturally look unfavorable. Initial expenditures are inevitably made to build staff. Tremendous bureaucratic energy will be expended on sorting the programs out. Infighting in the government requires issue papers and meetings, inefficient activity which does not translate into the delivery of services. At each level—federal, state, and local—debate will be required until there is finally a consolidation.

The third factor which policy-makers must consider is that there are significant measurement problems to demonstrating energy conservation results even if these programs could be shown to be effective in delivering services. For those state and local governments where it is possible to solve the staffing problem quickly, it will nevertheless be difficult to show that energy savings that would not have been attained otherwise will take place as a result of these programs. Even to show that there are any savings at all requires showing that consumption would be higher without the programs—an unprovable assertion.

One argument often advanced for building energy conservation capability in state and local governments—and one which must be considered by policy-makers—is that these governments are sometimes the cause of problems in implementing energy conservation practices. For example, the impact of local building codes can be significant. Even progressive communities frequently have commercial codes that do not require any insulation of buildings. Where utility costs are customarily passed through to the consumer or renter, landlords have little incentive to invest in energy conservation measures. Hence, encouraging state and local governments to become involved in energy conservation matters is a means for placing responsibility for taking action at the appropriate level of government. Even this accomplishment is difficult to measure.

A final factor for policy-makers to consider is that building more effective grant programs requires developing more accurate and more complete information. This seems simple. But there is little current understanding of who are the real consumers of conservation programs and what kinds of information it is that they need. Does a township need assistance in identifying where it can save money? Or, does the procurement office need help in designing a system that values life cycle costs? Does the homeowner not know what to do, or is he or she really concerned about what kind of insulation to buy and where to get it? Or, more fundamentally, do consumers need to be sure that they should invest in energy conservation because oil prices will remain at current levels and perhaps increase?

Coming to understand the varied needs for energy conservation assistance and learning what works best is the hardest problem of all. Developing more useful energy conservation services and improving delivery mechanisms require investments in experimentation. The need to rationalize the grant programs may be in conflict with the more basic experimental need to find out what programs work and are needed. Future budget debates will confront this issue which raises questions such as which programs should be judged by policy-makers to be cost-effective tools for encouraging energy conservation when compared with alternatives. What resources should they receive?

This chapter has discussed the philosophical problems that will continue to exist for some time in measuring the energy savings benefits of the grant programs. The discussion does not mean that they have no benefits, rather that quantifying them is misleading. For example, there are many anecdotes which illustrate that some programs seem to work very well. The problem comes in generalizing successful experiments to a national scale.

If the clearest benefits to be derived from the conservation grants programs to date have been shown to be in the support of innovation, then the pure approach would be to make some programs at least into research demonstration programs. Instead of encouraging the states to view these grants as "giveaways" while the most effective approaches are still being investigated, states might be required to apply for support. Only the most innovative or demonstrably successful programs would be funded. But as the constituencies that support funding for these grant programs mature, the opportunity to make them into large-scale research projects will become more difficult. Moreover, it is not clear, based upon an experience of other large-scale social research programs, that this condition is entirely unfortunate. In an atmosphere of continuing uncertainty where the most effective approach is still being sought, some critics of the grant programs will suggest that more structure should be imposed unless rigorous criteria are developed so that only those with the most likely chance of succeeding are funded.

Unfortunately, even the proven programs cannot be demonstrated at this time to work under all circumstances since, as we have seen, local factors such as outstanding leadership or community support have been particularly significant contributors to successful programs; imposing structure does not yet appear to be any more likely to make the grants programs cost-effective. In light of these circumstances, the best policy alternative appears to be a middle course. Flexibility should be retained, but the programs should be funded slowly. Innovation should be encouraged. Successful programs should be supported. If the hundred flowers are to be permitted to bloom, they should be nurtured with care.

Clearly, the grant programs should not be continued forever if they are unable to produce the energy savings that are expected of them. But at least until the nation better understands what works, a number of programs will be necessary if for no other reason than to encourage innovation and develop better evaluation techniques. There is at present no combination of services that is right for saving energy everywhere. Each state and local government may find slightly different approaches to be most helpful in solving their individual problems.

Roger W. Sant and
John C. Sawhill

Conclusion

The various chapters in this book have described the opportunities available for improving the energy efficiency of the American economy. The potential improvements in each major sector are striking and suggest that there is a real opportunity to decouple the growth in energy consumption from growth in GNP. William Hogan and Robert Pindyck, in their chapters, indicate that decoupling is a theoretical possibility. The chapters on transportation, residential and commercial buildings, industry and agriculture, and electricity generation and usage provide further support by showing how technology can lead to greater efficiency.

The pattern of U.S. energy consumption since the embargo also suggests that energy consumption can grow more slowly than the economy. During the twenty years preceding the 1973 oil embargo, growth in energy consumption and growth in economic activity were roughly parallel. In 1970, the U.S. consumed energy at a level of 62.5 thousand Btu's per dollar of GNP (1972 dollars) which was an all time high for the 1950–1970 period. Just seven years later, this figure had fallen to a post-World War II low of 56.8 thousand Btu's per dollar of GNP—a decline of 9 percent. For the 1970–77 period, energy requirements grew at roughly half the rate of GNP. Growth in GNP averaged 3.17 percent while growth in energy consumption averaged 1.75 percent or an energy-

---

ROGER W. SANT *is director of the Energy Productivity Center of the Carnegie-Mellon Institute of Research. Before joining the Center, Mr. Sant was assistant administrator at the U.S. Federal Energy Administration. A private energy consultant to several major corporations, he has written numerous articles and editorials on various aspects of energy.*

GNP ratio of 0.55. In the latter half of this period (1974–77), the ratio dropped to 0.47 with a 3.18 percent GNP growth requiring only a 1.5 percent annual growth in annual energy consumption.

It is too early to say whether these trends will continue, but, as the authors in this book have pointed out, the signs are very encouraging. The industrial sector, for example, has exceeded almost all prior expectations. According to data developed in the Department of Energy's Voluntary Business Energy Conservation Program, the energy productivity of industries monitored increased by 9.2 percent from 1972 through the second half of 1977. Robert Reid and Mel Chiogioji point out in their chapter that much of this improvement resulted from better housekeeping rather than major new capital investments in more energy efficient equipment; but there are some indications that new capital investment in more energy efficient technology is beginning to accelerate. One dramatic sign is that the overall energy efficiency of the industrial sector improved by 13.3 percent in 1977.

The reports from the residential sector are also encouraging. According to the American Gas Association, gas space heating requirements per customer per degree day were about 13 percent less in 1977 than the average usage in the 1967–72 period, and the decline appears to be continuing. Furthermore, a preliminary survey by the American Institute of Architects appears to indicate that current building designs are 10 to 20 percent more efficient than existing buildings and that improvements in the range of 35 to 50 percent will soon be possible with better education.

Robert Hemphill, in the chapter on transportation, lists the various measures which have been taken to improve energy efficiency in this sector. The most striking improvement has occurred in the automobile, where the data show that the 1978 models averaged 19.6 miles per gallon (Environmental Protection Agency statistics), which represents a 44 percent improvement over the 1973 fleet average of 13.6 miles per gallon. These gains should continue as the efficiency standards for new models are increased and new engines and lighter weight materials are introduced by the manufacturers.

The overall evidence on the economy, then, as well as the individual sector analyses lend credence to the general view that energy consumption need not grow as rapidly as GNP. This point was recently underscored by the Demand and Conservation Panel of the Committee on Nuclear and Alternative Energy Systems (CONAES) of the National Academy of Sciences:

> The results indicate that, given time to respond to prices, regulations, and incentives, U.S. energy demand is very elastic. Consequently, a major slowdown in [energy] demand growth can be achieved simultaneously with signifi-

> cant economic growth by substituting technological sophistication for energy consumption.

The CONAES report points out that energy consumption in 2010 could technically be as much as 59 percent less per unit of GNP than at present. And while many would dispute this estimate as being either too high or too low, there is a growing consensus that significant savings can be achieved without the detrimental economic impacts that were commonly predicted in the immediate postembargo period.

The improvements in energy efficiency which have been discussed in this book have been accompanied by a very marked change in attitude on the part of Americans toward energy usage. In the early 1970s, for example, the dominant view was that there were "natural resource" limits to economic growth and that energy conservation was necessary to insure succeeding generations that they would have adequate resources to maintain a standard of living at least equal to our own. The motivation for using energy more efficiently—following a period of declining real energy prices—was based on ethical rather than economic considerations. Perhaps the most widely publicized proponent of this view was the book *The Limits to Growth,* which appeared in 1972. The authors of this volume argued that civilization was at a critical juncture and that

> man can still choose his limits and stop when he pleases by weakening some of the strong pressures that cause capital and population growth, or by instituting counterpressures, or both. . . . The alternative is to wait until the side effects of technology suppress growth themselves, or until problems arise that have no technical solutions. At any one of those points the choice of limits will be gone.

This same conclusion was reached by the economist Robert Heilbroner in his influential book *An Inquiry into the Human Prospect.* Heilbroner draws a pessimistic picture of the future for Western democracies and foresees a collapse of our ability to adapt to emerging environmental constraints. He argues that we may be forced to sacrifice our political traditions of freedom and democracy in order to maintain the thrust toward ever increasing economic growth. This gloomy verdict is based on his premise that Western democracies are unlikely to make the necessary adaptive changes in the present to forestall disaster in the future since so many of the threats to future well-being are remote in time (the next century) or abstract and difficult to understand ("the CO_2 problem"). Similar sentiments have been expressed by such diverse people as Rene Dubos, Lewis Mumford, Charles Reich, and Theodore Roszak.

The shock of the 1973–74 oil embargo changed the public's perception of the energy problem and of the need for energy conservation. Suddenly, the small band of people who had been urging energy conservation on ethical grounds was joined by a much larger constituency which argued that energy conservation was necessary to reduce America's dependence on Arab oil. The motivation of these new constituents was patriotic and aimed at maintaining America's position of world leadership rather than any concern with ethical or philosophical issues.

At the conclusion of the embargo, much of this "patriotic" concern was dissipated. As the lines at the gasoline stations disappeared, people began to question the very nature of the energy crisis and no longer viewed conservation as something equivalent to a wartime emergency. The embargo, however, did succeed in inducing some marked changes in behavior and attitudes. And, as a result, according to a recent study by Daniel Yankelovich, the experience of shortages began a process of rethinking and reevaluation of life styles and social values—a process that was then suddenly and abruptly arrested with the lifting of the embargo. However, the memory of the crisis left some enduring fears in its wake and laid the foundation for changes in the future. Yankelovich found that for a large segment of the American public, perhaps a majority, "there was a readiness evoked by the crisis to change and adapt, and a flexibility with regard to the future that is almost invariably underestimated by the so-called experts."

One clear legacy of the embargo was the realization that "low cost" energy would no longer be available. In the 1973–74 period, oil prices increased almost fourfold, and there is every indication that further increases—of at least moderate proportions—will occur in the future. Thus, there is now a new motivation for improving energy efficiency—the impact of higher energy costs on industry profits and consumer well-being. People are clearly concerned with this problem. Consumers in one county in Texas, for example, were asked recently whether they felt that energy prices were a burden. The answer was an overwhelming yes, and when these consumers were asked what, in fact, they had paid for fuel, they overestimated the price of their gas bills by 80 percent and their electric bills by 50 percent suggesting that they may be exaggerating the difficulties of dealing with higher cost energy. Two-thirds of the energy consumers recently sampled by the Gallup Organization indicated that they felt their energy bill would be the same or higher in the next year. And more than 50 percent of the homeowners polled in the same survey felt that energy prices were making it difficult for them to make ends meet.

It is interesting that even though a large majority of the population views higher energy prices as a problem, very few consumers indicate a real determination to make the investments necessary to save energy. Only 14 percent of those responding to a recent survey indicated that they would insulate their homes in the coming year. Roughly the same number indicated that they would dial down their thermostats. On further questioning, only half of those replying favorably would commit to action specifically targeted at saving energy.

The major question people have relates to the "reality" of the problem. Yankelovich finds that 64 percent still blame the oil companies; 57 percent blame the Arabs; and 51 percent name the government. The difficulty of getting the American people and their leaders to treat the energy situation seriously and deal with it effectively—a problem which was identified in the introduction to this volume—is still very real.

This skepticism is pitted against the opinions expressed broadly that a stronger signal is needed if the momentum, which is currently stimulating a number of energy conservation activities, is to continue, and we are to approach the efficiency levels outlined in the CONAES study. The authors in this book uniformly agree that such a signal should take the form of higher energy prices. No one expects the "free market" can do the whole job or that energy problems will somehow vanish if only the government gets out of the way and lets the market and the unfettered profit motive operate. The limitations of markets are clearly recognized as is the necessity for dealing with some policy issues such as income distribution, natural monopolies, or environmental problems outside of normal free market mechanisms. Nevertheless, the argument put forth by most of the writers in this volume is that there is a very strong case for placing as much of the burden for stimulating energy conservation on the market as possible.

The question which must be addressed, however, is what criteria should guide policy-makers in establishing energy price levels. Three possible options are discussed below.

1. *Allow energy prices to move up to world market levels.* Energy prices in the United States, and in certain other industrialized democracies as well, are currently below the prices which have been established in world markets. In the U.S., this situation results from price controls on natural gas and petroleum. Various schemes have been advanced for permitting price increases to occur gradually, and it is likely that some type of gradual movement of prices up to world market levels will occur within the next few years. But, as long as prices remain below market levels, there will be less incentive to develop and install efficient technologies which can lead to large energy savings. The case for the rapid movement of prices to world levels, therefore, appears to be a very strong one, and it is strengthened by the analysis presented in William Hogan's chapter that the impact on the economy of such increases would not be very large.

2. *Allow energy prices to reflect replacement costs.* It is not necessarily true that permitting U.S. petroleum and natural gas prices to move up to world levels would result in prices for energy products which reflect

replacement costs. Even with domestic oil and gas prices at world levels, electric utility rates, for example, might continue to be regulated according to the "average cost concept"—i.e., by setting prices at the average cost of generating electricity over a utility's entire capital plant rather than permitting prices to the user to reflect the incremental cost of building a new addition to plant. If energy is sold below its replacement cost, consumers' decisions to purchase energy will be made in a market where the true cost of replacing that energy is not reflected in the price. Therefore, consumers will use more than they otherwise would and resources will be misallocated. Denis Hayes cites an example in his analysis of hydroelectric power (Chapter 9) where electricity from a new plant costs ten times more than the average price now paid by consumers. The situation is particularly troubling if one accepts the high, long-run price elasticity estimates which have been suggested by Robert Pindyck in Chapter 2.

3. *Impose energy price increases beyond market levels to achieve socially beneficial energy savings.* Some would argue that energy prices should be higher than either world prices or replacement costs in order to bring about savings more rapidly than would be done in a "free market." For example, higher gasoline taxes and oil import fees might be added to world market prices in order to reduce the national security threat caused by the continued heavy dependence for its oil supplies by the United States on unstable foreign governments. It is also possible, as was pointed out in the chapter by Douglas Bauer and Alan Hirshberg, to reflect some of the environmental costs associated with energy usage in its price. For example, it might be desirable to impose a sulphur tax on utilities to encourage them to install the technology necessary to remove sulphur and other undesirable chemicals from utility emissions. This tax would in turn be passed along to consumers in the form of higher prices for electricity.

There are clearly strong arguments for moving domestic energy prices up to world market levels. There are equally strong arguments for implementing some type of replacement cost concept. It is true, however, that there are difficult administrative and technical problems associated with using replacement cost pricing in practice, and it may, therefore, be necessary to continue to refine and experiment with this concept before it will be practical for use by utility regulators. Questions still remain concerning what costs to include as replacement costs; whether excessive profits will be created as a result of implementing this concept, and, if so, who should benefit; whether certain classes of customers (e.g., industrial customers) should be singled out for higher prices or whether to apply them uniformly to all consumers; etc. Work on these issues should be accelerated since the case for replacement cost pricing is compelling.

Higher energy prices, however, are only one part of a comprehensive package of policies to stimulate energy conservation. In addition to higher prices, there are other areas where public policy can be an effective tool for improving energy efficiency. In general, these actions represent areas where public policy can supplement and enhance existing market forces so that energy efficiency is improved more quickly or more effectively than might otherwise be the case. For example, in the utility sector, it appears desirable to use some type of artificial pricing mechanism to reduce peak load demand and thereby lessen the need for additional plant capacity. The Bauer and Hirshberg chapter suggests several ways of doing this and reviews the results of some of the experimentation which has been conducted to date. Mandatory mileage standards, energy efficient standards for new buildings and appliances, better consumer information on conservation technology, and tax credits for insulation also fall into this category. Most of these measures appear to be worthwhile. The danger, however, is that they will lead to too much bureaucracy and red tape and for this reason be self defeating. Robert Reisner's chapter on the growth of the state grant programs points out some of the difficulties in creating "energy bureaucracies." His analysis suggests that it is desirable to weigh the costs associated with any new "nonmarket" regulatory scheme with the projected benefits. The presumption in each case should probably be to use market forces whenever possible and to limit these other types of regulatory mechanisms to areas where there is a clear indication that the market is not working as fast as would be desirable and where the costs of regulation can be minimized. Indeed, pricing and regulatory policies should be seen as complementary, not competitive, alternatives.

One final area for consideration by public policy-makers is research and development. John Gibbons has made a very convincing case for a greater expenditure of federal funds for research. He has also suggested some reordering of conservation R&D priorities. His work—as well as those of some of the other authors in this book—indicates that it would be useful to find out why people have some of the attitudes toward energy usage that they do and what might be done to change these attitudes. For, as Robert Hemphill points out, if we could induce people to adopt a more favorable attitude toward public transportation, carpooling, and vanpooling, we could achieve significant improvements in the efficiency of our transportation system. Similarly, changes in attitudes toward desirable temperatures in residential and commercial buildings could significantly change the energy requirements for the residential and commercial building sector. Many other examples could also be cited—attitudes toward single-family versus multifamily dwellings, attitudes about living in cities versus living in suburbs, attitudes toward leisure time activities. Yet we know very little about this area and are spending very little to learn more.

Gibbons also argues forcefully that government research should be focused on those activities which would not otherwise be accomplished in the private sector. In this regard, he suggests that proposals which are designed to take a completely new and fresh look toward achieving a given objective receive high priority. For example, incremental improvements to a furnace should be left to the private sector. Whole new ways of providing comfortable living temperatures could be the subject of a government funded research program. Finally, Gibbons makes a strong case for more spending for basic research. Clearly, it will be necessary to expand government funding in this area if the U.S. is to maintain its technological leadership and continue to improve the energy efficiency of its industry at a rapid rate.

One deficiency which several of the authors in this volume have pointed to is the lack of adequate data for analyzing energy efficiency. The potential benefits of improving efficiency through substitution, based on experiences in other countries, were described in the chapters by Lee Schipper and Robert Pindyck. Denis Hayes in Chapter 9 points to some specific examples where materials, capital, technology, and labor can be economically substituted for energy as prices rise. Yet, it is clear that our knowledge of how substitution works and what is possible is only in its infancy. Learning more about the ultimate demand for the services energy now provides should lead to a better understanding of the technological options for substituting efficient processes and systems for fuels and electricity.

The central conclusion which emerges from this book is that energy conservation is still an extremely fragmented and disparate activity. Although there appears to be unanimity on the importance of improving energy efficiency, there are still important differences among experts and the public in general concerning how to go about the task. To fully realize the conservation potential which has been outlined by the authors in this volume will challenge our ability to bring about change—particularly changes in organizational infrastructure. New contractors, suppliers, manufacturers, and services must come into existence. Financial and other regulations and standards must adapt to new conditions. Better data need to be collected and more sophisticated models of the patterns of energy use in society must be developed. Furthermore, we must continue to develop and commercialize new technologies for improving energy efficiency, and, most important of all, we must accept the reality of higher energy costs.

Yet, in spite of this need for more work, cost based prices, and better information, a new wave of optimism—and commitment—appears to be emerging from many quarters that these changes are possible, desirable, and necessary. The Yankelovich study found consumers willing to change behavior and to make modest sacrifices in order to achieve a stronger economy. The energy problem today is viewed as an opportunity to utilize natural resources more effectively and at the same time insure a higher standard of living. Furthermore, it is viewed as an opportunity to help those in developing countries to speed the growth of their economies. Over the next few years, we can, therefore, expect to see considerable change and confidently look forward to some significant improvements in the way in which we think about and utilize energy.

Index

The American Assembly

Columbia University

About The American Assembly

The American Assembly was established by Dwight D. Eisenhower at Columbia University in 1950. It holds nonpartisan meetings and publishes authoritative books to illuminate issues of United States policy.

An affiliate of Columbia, with offices in the Graduate School of Business, the Assembly is a national educational institution incorporated in the State of New York.

The Assembly seeks to provide information, stimulate discussion, and evoke independent conclusions in matters of vital public interest.

AMERICAN ASSEMBLY SESSIONS

At least two national programs are initiated each year. Authorities are retained to write background papers presenting essential data and defining the main issues in each subject.

A group of men and women representing a broad range of experience, competence, and American leadership meet for several days to discuss the Assembly topic and consider alternatives for national policy.

All Assemblies follow the same procedure. The background papers are sent to participants in advance of the Assembly. The Assembly meets in small groups for four or five lengthy periods. All groups use the same agenda. At the close of these informal sessions, participants adopt in plenary sessions a final report of findings and recommendations.

Regional, state, and local Assemblies are held following the national session at Arden House. Assemblies have also been held in England, Switzerland, Malaysia, Canada, the Caribbean, South America, Central America, the Philippines, and Japan. Over one hundred thirty institutions have co-sponsored one or more Assemblies.

ARDEN HOUSE

Home of the American Assembly and scene of the national sessions is Arden House, which was given to Columbia University in 1950 by W. Averell Harriman. E. Roland Harriman joined his brother in contributing toward adaptation of the property for conference purposes. The buildings and surrounding land, known as the Harriman Campus of Columbia University, are 50 miles north of New York City.

Arden House is a distinguished conference center. It is self-supporting and operates throughout the year for use by organizations with educational objectives.

The background papers for each Assembly are published in cloth and paperbound editions for use by individuals, libraries, businesses, public agencies, nongovernmental organizations, educational institutions, discussion and service groups. In this way the deliberations of Assembly sessions are continued and extended.

The subject of Assembly programs to date are:

1951___United States-Western Europe Relationships
1952___Inflation
1953___Economic Security for Americans
1954___The United States' Stake in the United Nations
___The Federal Government Service
1955___United States Agriculture
___The Forty-Eight States
1956___The Representation of the United States Abroad
___The United States and the Far East
1957___International Stability and Progress
___Atoms for Power
1958___The United States and Africa
___United States Monetary Policy
1959___Wages, Prices, Profits, and Productivity
___The United States and Latin America
1960___The Federal Government and Higher Education
___The Secretary of State
___Goals for Americans
1961___Arms Control: Issues for the Public
___Outer Space: Prospects for Man and Society
1962___Automation and Technological Change
___Cultural Affairs and Foreign Relations
1963___The Population Dilemma
___The United States and the Middle East
1964___The United States and Canada
___The Congress and America's Future
1965___The Courts, the Public, and the Law Explosion
___The United States and Japan
1966___State Legislatures in American Politics
___A World of Nuclear Powers?
___The United States and the Philippines
___Challenges to Collective Bargaining
1967___The United States and Eastern Europe

____Ombudsmen for American Government?
1968____Uses of the Seas
____Law in a Changing America
____Overcoming World Hunger
1969____Black Economic Development
____The States and the Urban Crisis
1970____The Health of Americans
____The United States and the Caribbean
1971____The Future of American Transportation
____Public Workers and Public Unions
1972____The Future of Foundations
____Prisoners in America
1973____The Worker and the Job
____Choosing the President
1974____The Good Earth of America
____On Understanding Art Museums
____Global Companies
1975____Law and the American Future
____Women and the American Economy
1976____Nuclear Power Controversy
____Jobs for Americans
____Capital for Productivity and Jobs
1977____The Ethics of Corporate Conduct
____The Performing Arts and American Society
1978____Running the American Corporation
____Race for the Presidency
____Energy Conservation and Public Policy
1979____The Integrity of the University
____Youth Employment

Second Editions, Revised:

1962____The United States and the Far East
1963____The United States and Latin America
____The United States and Africa
1964____United States Monetary Policy
1965____The Federal Government Service
____The Representation of the United States Abroad
1968____Cultural Affairs and Foreign Relations
____Outer Space: Prospects for Man and Society
1969____The Population Dilemma
1973____The Congress and America's Future
1975____The United States and Japan